AF768905

The Island

I will arise and go now, and go to Innisfree,
And a small cabin build there, of clay and wattles made;
Nine bean rows will I have there, a hive for the honeybee,
And live alone in the bee-loud glade.

And I shall have some peace there,
   for peace comes dropping slow,
Dropping from the veils of the morning
   to where the cricket sings;
There midnight's all a glimmer, and noon a purple glow,
And evening full of the linnet's wings.

I will arise and go now, for always night and day
I hear lake water lapping with low sounds by the shore;
While I stand on the roadway, or on the pavements grey,
I hear it in the deep heart's core.

W. B. YEATS

The Island

BY ROBERT RUSSELL

Author of "To Catch an Angel" and "An Act of Loving"

The Vanguard Press, Inc.
NEW YORK

Published simultaneously in Canada by
The Copp Clark Publishing Company, Ltd., Toronto

Library of Congress Catalogue Card Number: 72–83352
SBN 8149–0721–0
Manufactured in the United States of America
Designer: Ernst Reichl

The reader will understand that, so far, I have been a most fortunate traveler through my years. My luckiest day, however, was that on which Elisabeth Shaw and I were married. Had she not also shared my love of wind and water, the smell of rain, and the songs of birds, my dream of an island life would be gathering dust at the back of some top shelf. Since this book records our joint efforts to make that dream our reality,

these pages are dedicated to her.

Contents

Introduction ix
1 Getting There 3
2 Auntie Em 18
3 Pike Fishing by Bell 37
4 The Fiberglass 48
5 The Dock 67
6 Ben's Story 84
7 Guests 97
8 Air Guns 112
9 Captain Outrageous 120
10 The Pontoon 132
11 Middle Age 145
12 New Openings 162
13 To Shake the Past 182
14 The Boathouse 194
15 On Ice 209
16 Approaching Autumn 234
17 Miranda and Prospero 246
18 With Miranda Still 257
19 Autumn 271

Introduction

In a chilly Oxford lecture hall a great professor was addressing his silent and gowned audience.

"But if you are interested in the meaning of a work of art—meaning or significance—neither of them good words, but we don't have a good one—the meaning, say, of a picture—focus your attention not on the center where everything will be bathed in too much light but on its edges—that borderland where light melts into shadow and the shadow becomes darkness. If meanings are to be found at all, they will be found at those dim edges, the lines where the clarity of statement is framed by the darkness of oblivion."

Whatever else he said that day, I did not hear it. Memory had taken me back more than twenty years and across the ocean to my home and to a game my sister Hunnie and I used to play. We called it "No Stepping on the Carpet." Starting from the kitchen door, the object was to make a complete circuit of each of the downstairs rooms, ending up again at the kitchen door without ever having set foot upon a carpet. The carpets were for the grownups. That's what they walked on, where they stood, clinked glasses, laughed, talked, argued, the grounds where they conducted the strange, urgent business of their lives.

I crouched in the wide doorway of the parlor, having just navigated around the big room with its old upright piano and scatter rugs that acted as an entrance hall. That had been easy, but the parlor would be another matter. The soft, opulent blues of the big imitation oriental covered almost the entire floor, leaving only a pencil line of a path for me. I stared down at the bare, polished wood. Its shine seemed to come not from above, not from the sunlight that flooded down upon it from the large windows, but somehow from underneath, as if it rose from the very darkness itself. I traced that strip of bright darkness I would have to walk and tried to figure out my future course. To my right, its back against the wall, glowed the overstuffed, oversprung red plush sofa. That was where Sunny Schneider would sit every Sunday morning when he came around after church to see my father. As they smoked their cigars, Sunny would talk and talk and talk, his voice rising. He would lean forward, shift about, stand up, slap his thigh, and almost shout, "I tell you, Jim, those fellows over at Cit Hall are half asleep. They are. What they need is to have someone give them a damn good shake. That's what they need."

Disturbed by his vehemence, one Sunday I had sidled out into the dining room to where my mother was setting the table and, tugging at her sleeve, I had asked nervously, "What's the matter? What are they talking about?"

"Politics, just politics," she answered, and returned to the kitchen.

Now the red sofa and my father's easy chair were empty, but the silent room seemed to buzz with their presence and the pressure of their mysterious concerns. I must try to skirt them. To do that, I would have to leap over a two-foot patch where the oriental met the wall to the big, soft armrest of the sofa. Once safely there, I could crawl to the other end, where another, but easier leap should bring me to a large expanse of wood glistening under a front window. If I could only reach that pool of shining, I could rest for a bit and work out how to get past the fireplace on the opposite wall.

"Come on, Rob! Come on. I'm right behind you," said Hunnie, just having cleared the carpet in the hall behind me. "You got to jump. Go on."

Taking a deep breath, I tensed my muscles and sprang for the sofa. My knees landed on the sloping armrest, and as they began to slide, I clawed desperately for a handhold on the smooth plush of the back.

"Haha! You touched. You touched," Hunnie jeered. "I saw you. You touched. Get out of the way. Now it's my turn."

Angry, sullen, ashamed, I walked slowly out into the middle of the room and turned to watch her effort. She crouched, sprang, and failed as I had done.

It may be more than coincidence that, decades later, we each have summer homes on islands in the St. Lawrence River. The river is not wilderness. We are not beyond, but on the edges of modern life. For me, that strip of shining wood has become a great and wonderful river. That river is now the path along which I thread my years and where, by the rocks and waves and wind, I search for my meanings.

The Island

1

Getting There

I was born in Binghamton, New York, in 1924 and blinded at five in a freak back-yard accident. For the next ten years I went to a school for the blind in New York City two hundred miles away, far enough away to mean that I could spend only the Christmas, Easter, and summer vacations at home. I hated New York. Its obvious failing was that it wasn't Binghamton. It didn't sound like Binghamton, smell like Binghamton, feel like Binghamton, and the food at school didn't taste like Binghamton. What's more, so far as I could tell, New York showed no signs of even wanting to become like Binghamton. Binghamton was just "country," as far as New York was concerned, so I decided I must love "country." That meant fields, woods, water, and the creatures that inhabited them; so I was an eager fan of the animal stories of Thorton Burgess, James Oliver Curwood, and Jack London. I felt sure they would have hated New York too, and so would George and Steve, my two Binghamton friends.

Through those summer vacations, the three of us wandered the Susquehanna and Chenango Rivers, sometimes with slingshots and beebee guns, but always with our fishing rods. There were patches of the Chenango's stony bed that my feet knew as well as they did our kitchen floor.

When dreaded September came, I took back to New York the feel of the rocks under my feet, the swirl of the current against my legs, the sun on the back of my neck, river smells, birdsong, and the taste of fresh corn snatched from a nearby field and roasted in the ashes of our fire on the bank. These memories were the bars of sanity to which I clung in those long, wakeful nights through which I lay listening to the intermittent thunder of the White Plains Express seven blocks away.

My thirteenth summer still glows, for it was then that I met the St. Lawrence. George and his parents invited me for a week's camping and fishing near Alexandria Bay over on the American side. As we set up our tent, I felt for the first time that big, soft wind and I could hear the river slapping at the boulders along the shore. Wind and waves spoke to my blood of size and power, great power, power that was not being exercised, but power that could be. This "could" was not a threat but the power was there. I remembered the time I shook hands with a light-heavyweight boxing champion. He did not, for him, "squeeze." His grip had not actually hurt, but my hand had tingled for days at the contact, the flesh knowing the strength within and behind what had enclosed it. But with all its wonder, his had been only the strength of a man; this wind, this water in motion—theirs was the musculature of the universe. It was not only my hand they held, but me.

"Will it squeeze?" I wondered as I lay on the hay we had mounded for beds, "because, if it does. . . ."

A numb, troubled adolescent, I listened to that wind booming in our canvas and those waves casually handling the rocks. Each moment was half filled with the fear that that great hand in which I lay might close, and half filled with joy that it had not.

In that night of wondering, I was alive in a way and to a degree new to me because I had the sense that every moment could be my last. For the first time I felt the presence of the ultimate danger. Every moment seemed a fresh, sweet surprise for which I had hoped, but to which I had no actual right. Those moments poured down upon and through me, a stream of gifts from timelessness. Each moment was there to taste, to be sucked dry, because I knew that no single one of them would come again and that the stream might, at any instant, cease to flow; once stopped, it would not start again. I have heard mountain climbers say they were most alive when each step could be their last. When they took that step and lived to take the next, theirs had been the courage to challenge death, so to them belonged the joy of having triumphed over him. Though eventually they would have to hand them in, their score cards would carry the lists of personal victories before the last defeat. Lying on the hay under warm blankets, I was in no actual danger, but I felt that I was, and that is what mattered. Sensing that danger, to be alive was inexpressibly sweet. Next morning we would go out onto the river rolling and tossing under the wind—no small, tame river like the Chenango, to be walked out into with bait box and rod and whose deeper stretches could be fished from a leaky, flat-bottomed rowboat for fifty cents a day. The St. Lawrence was another world, and all through that first night I lay at its edge, wondering whether I would have the courage to enter upon it.

The sausage, beans, eggs, coffee, and even the big slabs of buttered bread tasted of the open fire. After breakfast, George and I took the mugs and tin plates down to the river to wash, and his father went off to rent a boat. My scalp prickled with excitement at the first touch of the cold water. From far out came a distant humming.

"Hey, what's that, George?"

"Couple of guys trolling," he answered.

"What! Way out there?" I asked. I tried to imagine how far out they must be for the distance so to muffle the sound.

"Yeah," was his casual reply. I wondered about their chances if something happened.

"They must be all the way over by Canada, eh?" I suggested.

"Heck, no. They're not even halfway. This is a big river."

"Must be," I mused; "sure *must* be." The size of the world through which we move, our sense of space, is determined by the limits of our perception. The eye grasps far more than the ear; so I was having trouble conceiving the size of the stage onto which I was shortly to move. It more than bothered me.

Every seafarer will remember the heart's sudden chill when for the first time he looked over the ship's rail and saw nothing but water. If disaster were to strike, which way ought he to swim? How far would he have to . . . how far could he? If I could have seen the green islands among which the fishermen were moving, the river would not have seemed limitless, but islands, however green and substantial, make no sound. Once out there on the river, for me there would be no land, only space through which wind and water would ripple and hiss. This was to be my maiden voyage whose edge of danger was keen, not yet blunted by the plastic coating of familiarity. But the sense of danger brought also the thrill of high excitement. All that was peaceful, tame, secure, comfortable—all that was behind me. I was setting forth into the strange, the new, the terrifying, the wonderful, and all my nerves twanged with the tension.

The rock I knelt on as I scrubbed the plates felt like coarse sandpaper, its surface diamonded with big crystals. And the rock had edges, sharp, angular, definite, strong. The rocks of the Chenango, I realized, had been smoothed, rounded, softened, dulled, and domesticated almost as if by centuries of civilization. The St. Lawrence rock was raw, strong, butting its shoulder up from the very foundation of things. And the water that chilled my hands felt clean, soft, sweet, as it moved past me from Lake Superior, fifteen hundred miles to the left, toward the Atlantic, a thousand miles to the right. And I was going out there, out upon it!

"Hurray!" shouted George, rising, "here comes Dad with the boat!"

"Is that him?" I asked, motioning to the left where an outboard seemed putt-putting toward us.

"That's our old Muncie," George almost sang, "all five horsepower of her."

"What's it like?" I asked nervously. "What's the boat like?"

"A flatbottom," he answered.

"But how big?" I pressed.

"Big as a scow," he laughed happily.

"Does it look safe?" I continued. "Flat-bottomed and all, I mean?"

"Well, it's *big* enough," he evaded. As it chugged toward us, I went over all the arguments I could remember about the merits of flat- as opposed to round-bottomed boats. The pitch of the motor dropped to a series of more or less regular coughs.

"He's bringing her in," George cried. "Come on; let's get our stuff."

Up the bank, back to the camp, then down to the river again laden with rods, nets, tackleboxes, bags of sandwiches, and whatnot. Knees shaking and heart pounding, I waited while George and his father stowed the gear.

"All right, Bob," said George's father, "come aboard. Give him a hand there, George." Half of me—maybe two-thirds—didn't want a hand, was afraid of the hand, was afraid to go, but the remaining third was too ashamed to admit that he traveled in such cowardly company, so both of me climbed in and sat down. George had an oar and was shoving off.

"Haven't we forgotten anything?" I muttered weakly, but they weren't listening.

"Atta boy," George's father congratulated him, yanking the starting cord, and we were off.

The next seven days were full of the slapping of waves, rocking of the boat, flopping of fish, the feel of sun and wind, gusts of rain, and evenings with the smell of wood smoke, frying bass, hay, and the sound of serious voices over the campfire as the others examined the chart of our part of the river, trying to decide where the bass might be next day: over by Whiskey Shoals, Gull Rock, or Tibbetts Light. Binghamton might still be somewhere back there or it might not; it didn't matter. Nothing mattered except what I was doing, hearing, smelling, tasting, being. The present was all there was, and it was more than enough. The details of those days are still bright, but the dinner we had at a roadside restaurant on the way back lies half in shadow, and the arrival home is lost in the darkness of normal living. That was my first meeting with the river.

That summer I learned the lesson Sophocles had learned centuries earlier by the Aegean about the on-goingness of things, about how time cannot be stopped even though the days be golden. For, despite my having met the river, September came anyway, and wearily I boarded the train once more for New York. If I hated it earlier, it seemed now intolerable, now that I knew about the St. Lawrence. Grimly I determined to leave that place as soon as possible and quite consciously set about compressing the usual twelve years of school into ten. At sixteen, I was a freshman in college. I no more boast of this than a man boasts about breaking a world record when chased by a lion. The escape is what mattered, and I escaped from New York to college where emotionally I found in literature what physically I had found on the river.

With a B.A. and M.A. in English from Yale, I lucked into my first teaching job at what is now the State University of New York at Binghamton. It wasn't as if they had wanted me—that is, they didn't want a blind instructor because, having never known of one, they couldn't believe I could possibly "work out." Necessity forced their hand. Colleges were bursting in 1946 with the flood of returning GI's, but instructors were scarcer than warts on a duck's back. In this particular English Department, scheduled classes had already been meeting for two weeks without an instructor. Figuring, I suppose, that even *I* would be an advance over the empty chair up front, the Dean bowed as gracefully as he could and hired me. Apparently I "worked out," because the next spring the Dean rehired me with a generous raise. He apologized profusely about his previous misgivings. I told him that was okay because I had had doubts even more serious than his.

In my second year there, I realized that, if I were going to continue in college teaching, I would have to get a doctorate. My office mate was an Englishman who planted and nourished the idea of Oxford in my head. I applied for a Rotary International Fellowship for graduate study abroad, and it came through. While I had come to like the people I worked with, it gave me special pleasure to hand back unsigned the contract the Dean offered me for a third year. In just two years I had risen from the misery and frustration of joblessness to what seemed the pinnacle of success. Instead of hunting desperately for a job, I was turning one down, and how sweet was its taste on what had been a parched tongue!

When September came in 1948, I was not boarding the train for New York, but the *Queen Mary,* bound for Southampton. If Sophocles knew that time could bring joy as well as grief, he didn't say much about it, which is a shame because that's just as true as the reverse. Two successive Fulbright Scholarships followed the Rotary grant, so I had three years to learn the Oxford libraries and its almost-as-famous pubs. It was in a library, though, that I met Elisabeth. She was one of a whole troop of women undergraduates whom a friend had organized as volunteer readers for me. She was doing English too, and in the year after she took her degree—my last at the university—we were married. Most of her childhood had been spent on Orkney, an island off the north coast of Scotland, then another ten years in sea-hammered Cornwall.

"In your country we will live near the water, won't we?" she inquired.

"Of course. I promise."

My first teaching job back in the States was in northern Illinois. Our horizons were bounded by thousands of miles of cornfields and a salary of thirty-five hundred. The Mississippi, cold, muddy, dreary, trickling along ten miles to the west, didn't count.

"But," she kept saying, "we've got to get out of here!" and I kept saying, "I'm *trying!*"

I was. I wrote to every college and university in the United States. I could have papered our apartment with their rejections if I had known how to paper. But in my third year I struck it rich—forty-one-hundred dollars' worth—at Franklin and Marshall College in Lancaster, Pennsylvania. If I had enjoyed resigning from that first job in Binghamton, it was nothing to what I felt handing back the contract in northern Illinois. A kind friend read the last batch of my students' exams to me as Elisabeth packed the car in June.

"Give him an A," I said, when she had finished the last one, "and thanks."

Elisabeth had the car running and, with Richie, three, and Mark, five months, stowed in the back seat, we headed east.

We sweltered through the last three weeks of June, settling into our Lancaster house.

"But where's the water you were talking about?" Elisabeth gasped. "You said there was water in Lancaster."

"Not *in* Lancaster," I corrected, "not actually in the town, but there's the Susquehanna only a few miles down there, and the Bay . . . Chesapeake Bay is somewhere around here, I think."

"Then let's go," she said. "I've had about enough of this heat."

"Now, for water, if you want water—if you want water and cool weather—there's really only one place."

"Where?" she asked wearily.

"The St. Lawrence River."

"How far is it?"

"Two hundred, maybe three hundred miles."

"Let's go to the Susquehanna then," she said.

"Wait! We've got to think ahead. I mean, there are going to be more summers like this." Elisabeth groaned. "And we don't want to spend them in heat like this; so the only thing is to buy a cottage."

"With what?" she asked dryly.

"And the only place to have a cottage," I ignored, "is on the St. Lawrence."

"Then let's go."

We did, but it was four hundred, not three hundred miles.

We presented ourselves to real-estate people in Cape Vincent and Clayton, New York, declared that we wanted to check out their listings, and, to our surprise, they didn't ask whether we had any money. Instead, they showed us what they had along the shore and even took us out to a few places on islands. They were all either too expensive or in need of too much repair. When we had run through what was available, we took an afternoon to relax on the shore. Elisabeth held Mark and watched Richie paddling in the shallow water. I threw in a line.

"So beautiful," she said sadly.

"Crazy, I guess, to think we could ever have something up here," I said.

"But it would have been . . ."

"Hey, a *bite.*" I set the hook and felt the electric tension along the line.

"A big one?" she asked excitedly.

"Pretty good," I said. The fish leaped, and my line buzzed in the instant he tried to shake the hook. "A bass!" I cried.

"How can you tell?"

"They're the only ones that jump like that." Slowly I brought him in as he tired. When he was only a couple of feet from the end of my rod, I passed it back through my hands till I held the line. Working gently down it, I finally reached him. Sliding my fingers over his head, I closed them behind his gills, and I held the sleek, tense, beautiful form of a St. Lawrence bass.

"Hey, Richie," I called, "come here!" He splashed over. "Look at this." I held it up. The seventeen years peeled back. I was again in touch with the river.

"Fish," he said in wonder and admiration.

"A bass," I said. I cleaned and filetted him on the rocks, and we had him for supper.

As we sat on the steps of our rented cabin that night, Elisabeth gave words to both our thoughts.

"Suppose we ought to be getting back." The words lay in the warm, dark air like the breath of old approaching Septembers. I thought of the bass, the first touch of that big wind years ago, and the smell of hay.

"We could try Canada," I suggested. "We're close. Just over the bridge." CANADA! The name carried all the magic of northness—the whole progression from dark, Arctic blizzards down to where man might live, to where he could live, to where he did live on the hundred-mile strip along that country's southern border. To go "up there," though it meant only a few miles, would be an even deeper step into what the river had meant.

"How about trying that tomorrow? British air. They even speak English instead of American."

"I thought they spoke French."

Next morning the real-estate agent in Gananoque spoke either English or American, or maybe American English. Anyway, we understood each other.

"Sure," he said confidently. "You want to be on an island or the mainland?"

"An island," I answered immediately.

"Do we?" Elisabeth asked. "Why?"

"Well, on an island there's more . . ." I couldn't say "river," even though that's what I meant, so I said, "privacy."

"Come on, then," said the agent.

It wasn't till we clambered out of the boat at the first "property" that I re-examined my mental image of an "island." As we cautiously picked our way up a semicliff, I realized that I had always thought of an island as a circular plot of flat land—definitely flat, or at least with no more than one or two gently sloping mounds perhaps three feet high—carpeted with moderately tall grass and shaded not necessarily by palm trees, but with trees that unmistakably leaned in that direction.

"Careful, Bob," Elisabeth cautioned, steering me to the right, "here's a crack, a crevass, my God! it's a . . ."

"The view," encouraged the agent. "The view . . . that is, once we get to the top. Now, that's really something!" I thought about that.

"It's all just rock, Bob; just *all rock,*" Elisabeth said doubtfully. "No trees; no land, really. Just rock!"

"She won't wash away," chuckled the agent, "no sirree! Now, here's the cabins right up where you get the view."

At the top, two small, one-room cabins faced each other across ten feet of flat granite.

"Lord!" Elisabeth moaned, "it drops straight down over here! That must be sixty feet. If you came out of one of these cabins and turned the wrong way . . ." I thought of numbering the cabins One and Two. Coming out of One, I turned right; coming out of Two, left. All I would have to do would be to remember which I was in—that is, not forget.

"A fella could put up a rail," suggested the agent. "No trouble at all."

"But the children?" wondered Elisabeth, "where would they play?"

"Well, you've got this nice patio right here," he stamped on the rock, "for outdoors, and when they got tired of that, well, they could go inside. Why, they could have their own cabin all to themselves, and you could have yours. Ideal setup, eh? Yessir, ideal!"

"What about water?" I inquired.

"Oh, plenty of that. Plenty. No worry at all. Deep water all around. Any kind of boat could pull right up to the dock."

"No," I countered, "I mean up *here*. Water to drink . . . wash . . . things like that."

"That's one thing there's no shortage of," he explained good-humoredly; "couple of buckets, that's all."

"Yes," answered Elisabeth. "It's getting them filled." He laughed.

"You said you had several places?" I said.

"Sure," he replied, and we started down the goat track.

"If we should be interested," I said, "where would we get in touch with the owner to ask . . . well, other questions?"

"Oh, he's down in the States somewhere. Isn't up here very much. Makes him nervous or something, he says. Getting a little on in years, he is. Watch your step there."

Next came a pair of small, ancient, but comfortable cottages. The owner was not too nervous to be there.

"We live here in the larger one," she said, "and the smaller one is for guests. A very nice arrangement. We have our privacy, and they do too."

"Ideal," chimed in the agent. "Plenty of room for the kids to run around . . . hundred acres or so. Course, this one here is only a small lot, but you've got your frontage, two cottages, and some good trees, plenty of neighbors . . . well, not too many, not too many . . . maybe seven or eight, and so on like that. Perfect spot." The land was flat; the grass was only moderately high, and there were trees.

"How about the fishing?" I asked.

"Marvelous," said the woman. "My husband's out now. He's a great fisherman. Pike, bass, everything right around here."

"And you're just a two-minute run from the Forty Acres," added the agent.

"What's the Forty Acres?" I inquired.

"Why, that's a patch of shoals out in the Wide Waters where they catch all the muskies. Most famous fishing spot in the whole St. Lawrence, and just outside your front door, as you might say. And all for thirty-five hundred."

Obviously, he was right. It was ideal. From the brain waves I thought I was receiving from Elisabeth, she felt the same way. Still, we couldn't actually come out and say so without a conference, without a little serious soul-searching before saying we would take it. Apart from anything else, it wouldn't be fair to be that casual with someone else's money, even if that someone was only a bank.

We munched our butterless cheese sandwiches on the benches by Gananoque's town dock.

"Well?" I said.

"But the money? Where are we going to get the money?"

"The question is: Do you like it?"

"Beautiful," she replied. "Much the best of anything we've seen. It's . . . well . . . our sort of place, and . . ." We went on reciting all the ways in which it was "ideal." "But the money?" she ended. "Should we call the bank and make sure before we say we'll take it?" I imagined myself in a telephone booth pinching each coin that I dropped into the slot, saying, "Hello, Bank? This is Robert Russell calling. I've got a job and I've just found the perfect cottage. All I need is thirty-five hundred dollars. Okay?"

"No point in calling the bank. Let's tell the agent." Shakily, I told him.

"I see," he said, "I see. And how much . . . just how much would you like to put down? To secure the deal, you know."

"Well . . . nothing, right at this moment," I admitted. "I'll make . . . make the arrangements with my bank as soon as I get back, but I'd like to settle things before we leave so we can rest easy."

"Sure, sure," he said. "Sure, sure." After a troubled silence, he rose. "Well, let's go over to the island and see what they say."

As we buzzed back over the mile and a half of river, the air smelled like no air I had ever breathed. It had the inexpressibly sweet tang of possession—our air.

"I'll just go up and explain things," said the agent as we docked. "No need for you folks to come." As our wind breathed through our trees and our waves slapped on our rocks, my thoughts reached down the shining corridor of summers to come.

"That's what having a family means," I observed, "having your own place."

"And *such* a place," she answered dreamily. "*Such* a place!"

"Sorry, folks," said the agent, coming down to the dock. He stepped into the boat where he sat for a moment. "The thing is," he said, "but the thing is . . . I know it sounds crazy, but the thing is that it's sold."

"It's *what?*" we chorused.

"It's sold," he repeated.

"But we just . . ."

"I know," he continued, "it was for sale an hour ago, but it's sold now. One of the neighbors," he explained. "They just came over and bought it after we left."

"But why? How could they?" I groaned.

"I know," he commiserated. "It was for sale all summer and they didn't buy it, but they've bought it now."

"It's gone, then?" Elisabeth wept. "Really *gone?*"

"That's it," he said, starting the engine. "Guess we'd better be getting back to the mainland."

The mainland. That's where we went the next day—all the way back—and it felt like it, all four hundred miles and those long months that followed. Every few weeks that winter one of us would say to the other, "I don't suppose the bank would have given us the money anyway, so what's the use?"

"Probably not. Why should they?"

But one sunny day next April something got through to me.

"But they might," I said after dinner.

"Who might what?" Elisabeth asked.

"They might lend us the money."

"Never," she answered. "But anyway, it's sold."

"There could be others. Not exactly like it, but there might be another."

"Then write and ask, but it isn't likely. Nope. We've missed it."

I wrote to the agent in Gananoque and asked and waited.

"He says there is one," Elisabeth greeted me from the top step as I returned from the college, and she read his letter to me there on the porch. "Bigger, better, and cheaper. On the same island, Hay Island. Two doors down from the one you almost bought. Four bedrooms, large living room, fine fireplace, plenty of room for the kids. All you've got to do is pick up your groceries and move in. Ideal. And only twenty-five hundred." I felt the wind and heard the water.

"But here it is, only April," Elisabeth said sadly, "and we couldn't even get up to see it before June. It'll certainly be gone by then."

"Maybe, but maybe not."

"Maybe? A buy like that?"

The river formed the main current of my thoughts, and I talked about it anywhere and everywhere. I bent the ear of a psychologist in the coffee shop one morning. He was a fisherman, so I knew the ear would be sympathetic.

"It's only a dream," I ended, "but if it were only possible. . . ."

"Possible!" he echoed. "What a word!"

"What do you mean?" I asked.

"Well, just how much of our lives," he continued ponderously, "just exactly how much of our lives is bounded by that word? How much? And who gives it meaning? Who? Why, we do. We say that something's 'not possible,' so we don't do it. We decide beforehand that we can't have whatever it is, so we never try, and we never get it. It's crazy."

"You think it *is* possible, then?" I asked.

"Why not?" he said.

"How about you?" I countered. "Wouldn't you like a place like that?"

"Sure."

"Then why don't *you* go buy it?" He was silent. "Well, why don't you?" I pressed.

"Too complicated," he said, getting up. "Anyway, I've got a class."

I sat there in the clatter and buzz of the cafeteria with my empty cup, wondering. The thing was, I decided, that he didn't actually want a place like that. That's why *he* didn't go buy it. He didn't want it, but I did. After dinner that night, I said: "How about picking me up after class tomorrow? I'd like to go to the bank."

"What for?"—as if she didn't know.

"Won't hurt to find out," I said. "No harm in that."

"But we don't even know what it's like!" she argued.

"We'll find that out too," I said. "Anyway, we do know. We've been there."

"But not the house," she urged, "and the house is what we're buying."

"Not really," I said. "Anyway, we'll find out."

"We'll certainly consider this application, Mr. Russell, at our next loan-committee meeting, and we will be in touch. Thank you for coming in, and personally . . . personally . . . Canada, you say?" His voice lost its official edge and went wandering slowly off among old dreams. "I've always thought . . . well, I used to think . . . I guess everybody thinks about a place like that. But," he said, snapping back to his position behind the paper-piled desk, "we'll be in touch. Good-by."

Though it took a week, he was in touch. The bank officially approved.

"They're crazy!" Elisabeth cried. "If we had any money in our account with that outfit, I'd cancel it."

"Look," I said to the same psychologist the next morning over coffee, "I'm still going on with this business, and you're a fisherman. How about snatching a weekend early in May and driving up to look at this place I was talking about?"

He considered. "Now *that* would be possible," he said. And we did.

The signing of official and meaningless papers, the handing over of checks, the receiving of deeds, and all that sort of thing is, to be sure, a critical part of purchasing property. Though this was my first time through that dignified and formal ceremony, I can't remember when it happened. However and whenever it happened—the event legally certified me as an islander of the river. In actuality, I had become one seventeen years earlier, but had only just gone through the formalities of taking up physical residence.

2

Auntie Em

"Of course there are a few things that need doing," I told Elisabeth on my return. "I mean, you can move right in with the groceries like the real-estate agent said, but there actually are a few things that must be taken care of."

"Like what?" she inquired suspiciously.

"Well, for example, I couldn't get the water pump to work—couldn't get the engine started."

"Engine? What engine? What do you mean by engine?"

"There's a gasoline engine down beside the sink. . . ."

"In the kitchen?"

"Sure in the kitchen, and you pull it like an outboard."

"What about fumes? Aren't they poisonous?" she pressed.

"There weren't any," I said.

"But you didn't get it started. I mean, when you do start it, what about the fumes?"

"You just open the door, that's all. It's right by the back door."

"Mm!" she pondered. "And what else?"

"Oh, little things here and there. The problem is that they're going to take money to fix. That place can be put in apple-pie order for just a few hundred. The agent said so himself, and there are plenty of handy fellows up there who work cheap."

"And where's that few hundred supposed to come from?" she asked pointedly.

"Teach summer school . . . the only way . . . just the first session . . . to pick up the cash. . . ."

"And you think the rest of us are going up there without you?" she exploded.

"I don't see any other way. I mean, if I could fix a roof, I would, but even then there's the lumber. . . ."

"Roof? What's wrong with the roof?"

"How do I know? I didn't crawl out there. He just said it could be fixed up for a hundred or two, and good Lord, the place was cheap enough."

We had to go over this ground several times before Elisabeth saw the basic soundness of my position. Finally I clinched it with, "It isn't as if I won't want to be there, you know. I'm going to be miserable, absolutely miserable down here. Remember the heat last summer? And you'll be up there on the water. Besides, you're not going to be alone. You'll have Richie, Mark, and there'll be Auntie Em." Auntie Em was Elisabeth's great-aunt in her middle seventies, "an English maiden lady," as she described herself. She was coming to spend the summer with us. "And she'll be a great help to you and good company." That settled it. Elisabeth went and wrote letters. I stayed and was even more miserable than I had promised. The heat was stifling, the students lazy, and I taught badly. As I lectured, I couldn't get the sound of water out of my head, and then there were her letters. When that first session dragged to an end in the middle of July, I seized my four hundred dollars and took the next bus north.

Fifteen hours later, I kissed Elisabeth in the Texaco station that served as Gananoque's bus stop.

"You look tired, darling," she said. "Hope you're not too tired to row, because that damn motor doesn't work half the time."

"How's the boat, though?" I asked.

"It doesn't leak, if that's what you mean, but it's too small! It's just too *small.* Everybody up here says we're crazy. That boat was made for one cautious adult on a millpond, and this is no millpond, I can tell you."

It was a featherweight car-top rowboat I had bought during the winter before we had heard about the availability of the cottage. I had answered an ad in the paper, and it looked just great there in this fellow's garage; so I snapped it up on the grounds that, since we were definitely going to buy a cottage, we would certainly need a boat. I brought it home and hung it in our garage, where I used to go out to stroke it when things seemed especially bad. I had also nailed a great buy on a fifteenth-hand, five-horse motor. It ran like a dream in the testing tank.

"We'll manage all right," I said cheerfully. I had to row only the last half mile.

"Careful. This dock is very high. Dick Johnson and his father have been terribly nice about it—about everything, really—they said we were very welcome to use their dock since we don't have one." Climbing out, I made the bow fast and Elisabeth tied the stern.

"Where does our land begin?" I asked as we reached the shore.

"Just between these two cedars," she answered. "Soon as we hit the tall grass." And it was tall. "I found an old scythe in the shed and tried to cut a path at least down to the river, but I guess it's dull. Mind! Here's one of those bloody thorn trees. They're all over the show. We'll have to do something about them. Can you sharpen scythes?"

I didn't know what a scythe looked like, but I said I could sharpen ours.

"And we have the loveliest willow—biggest I've ever seen—right down by the water. *Children,*" she called, "here's Dad!" and Richie and Mark came thumping down the front steps. As I hugged them, they felt like two little mattresses.

"I don't let them out of the house without life jackets," she explained. "They're so used to them they don't even know they've got them on." Auntie Em had a pot of good, steaming British tea ready for us.

Despite all the happy chatter over tea as everyone told me what it had been like—about the storm, about the carpenters who had fixed the roof, about the genius who had finally coaxed the water pump into action—depression settled more and more heavily upon me.

"What's the matter, darling?" Elisabeth asked as we went to bed.

"Just tired, I guess," I replied, but that wasn't it. The trouble was that I was only a guest. The place was theirs, not mine, not *ours.* They were the ones who knew it, had lived in it. They had struggled to clean it up, made the arrangements with the neighbors about the dock, hired the men to repair the roof, decided about wearing life jackets. They had their routine and were coping with things. All I had been doing was making the money. I promised myself that, however bad things got, I wouldn't let that happen again. Roof or dock, thorn trees or water pump, I wanted to be there. Just making a living was no good compared with making a way of life.

"Now, as to bats," said Auntie Em next morning as we sat around the breakfast table over our third pot of tea, "what exactly can you tell us about them?"

"Well . . . not that much, actually," I confessed. "In fact, I bet *you* could tell *me* something. I know we have a few."

"A *few!*" cried Elisabeth. "We've got colonies!"

"Yes, dear," said Auntie Em, "I expect we wouldn't have quite so many if . . ."

"All right, all right," Elisabeth groaned, "but they did look awful. There were all these bits of paper. . . ."

"And rags," interposed Auntie Em.

"And rags," agreed Elisabeth, "*dirty* rags all jammed into cracks and knotholes in the wall . . . just all over the place."

"And Elisabeth in*sisted* on pulling them all out," Auntie Em declared. "She just couldn't leave them alone, Bob. Just absolutely couldn't. I told her they were there for a reason. They had to be. People don't go around sticking bits of oddments into walls just for sport, now do they, Bob? What! What! Well, I told you, didn't I, dear?"

"But they looked terrible," Elisabeth defended, "so naturally I pulled them out."

"Yes," said Auntie Em, eagerly leaning forward, "she did. And do you know what, Bob? Can you guess? Well, it's through those jolly old holes that the bats come." She ended in virtuous triumph.

"Do they?" I asked.

"Indeed they do," she replied with satisfaction, leaning back. "Indeed they do. Quite, quite. Through those exact same cracks. They seem quite happy in the attic all day, but they come down at night."

"You ought to see Auntie Em swing a badminton racket,"

laughed Elisabeth, "or a hat! I heard this terrible commotion one night, so I got up, and there was Auntie Em, standing in her night things, a candle in one hand and her hat in the other, swearing at them and swinging her hat."

"I don't swear, dear," snapped Auntie Em, "but those beggars just keep flapping around and flapping around, and you know, Bob, you never know where the little blighters are going to land! I don't think they know themselves, that's what I think. *What what?*"

"Have bats, do you?" said the clerk at the hardware store in town.

"That's right," I said, "and I'd like to get rid of them."

"Would you?" he chuckled. Then, after a pause, "Sulphur's the thing," he explained. "You burn these sulphur candles, and the smell drives them right out. Yessir, right out."

"Do you have any?" I asked.

"All kinds," he replied. "How many would you like?"

"Three . . . four . . . well, give me a dozen." He popped a bag and started counting candles into it.

"There we are," he said. "One thing, though," he commented after giving me my change, "you want to be kind of careful how you burn them. Put them in a bucket of sand or something."

"They flare up, do they?" I asked.

"Do they *ever!*" he exclaimed. "And fire's the main danger on the islands, you know. Not much you can do about it once it gets started. Just stand far enough back is about all."

"Got anything besides sulphur candles?" I inquired nervously.

"They're about the most effective," he observed thoughtfully, "but you got to be careful. Course you got to make arrangements beforehand."

"You mean with pails of water and so on?" I asked.

"Good idea, but I was thinking about having someplace to stay."

"Stay?"

"Yes," he answered, "because sulphur doesn't smell all that good. Fact, that's what drives the bats out. Guess it's poisonous too."

"You mean we'd have to move out?"

"I would if it was me," he replied. "Oh, you can come back, of course, but then, so do the bats. They do if they can."

"They come *back* after the sulphur and everything?"

"Well, look at it this way. It's their home too. I've heard they don't like the smell of some of those insecticides. They say you just pump it through the walls or whatever, and they leave."

"Now, that sounds more like it," I said. "No danger of fire with that treatment."

"Nope," he agreed. "Like a gallon, would you?" I said I would, and he bagged one plus another large chunk of my cash.

"Do they come back after the insecticide?" I queried.

"Pretty much the same," he admitted. "They can smell their droppings, they say, and they follow the smell right home. Oh, you can chase them out lots of different ways. Trick is to keep them from coming back."

"That's what I want to know," I said angrily.

"Guess plenty of people would like to know that," he chuckled. "Only way I know is to close up the cracks. They say a bat can squeeze through anywhere you can slide a dime."

"It's an old cottage," I said sorrowfully.

"Lots of cracks, eh? Probably got loose boards in the eaves. I know the sort of place. Paradise for them."

"Well. How do I close them up?"

"There *is* a way," he said confidentially. "Wire mesh. Wire mesh's your answer. Very small wire mesh."

"You mean . . . you mean I have to cover that whole house with wire mesh!"

"Do the job," he said simply. "Yessir, that'd do it."

"How much is . . . no! Never mind. Hell with it! I'll burn the bloody place down first."

"That'd get rid of them too," he said cheerily as I left.

"But if it's dangerous," Elisabeth exclaimed, "I mean, if those candles are dangerous, why risk it?"

"Bats!" I explained as I adjusted the rickety stepladder under the trap door in our second-floor ceiling that led to the attic. Climbing up, I raised the loose section of ceiling and was instantly engulfed in the warm, rich, brown smell of bats. No wonder they could find their way home, I thought.

"I think it's crazy," she said, "just plain crazy." I came down, picked up my bucket of sand in which nestled the potential conflagration, its stiff wick protruding like a fuse.

"Getting fond of bats?" I suggested, climbing back up with my bucket.

"Where exactly are we supposed to go when you light that thing? You said it was poisonous."

"I'm just . . . just placing it," I said. "I'll actually light it some morning when . . . when we've got things arranged."

"There's no floor up there," she warned. "Be careful. Walk on the studs or you'll come through the ceiling."

"They're not *studs!*" I snapped. "Studs go up walls. These are joists. Don't you remember that article about houses in *Consumer Reports?*"

"Whatever they are, there's no floor," she returned.

Standing on the top step of the ladder with my head and shoulders bathed in the warm bat-rich air, I listened for squeaks and the flap of leathery wings.

"Hear anything?" she asked nervously from below.

"It's all right," I said bravely, and I raised the heavy bucket and set it down gingerly between the joists beside me. Still no sound. Courageously I gave a little jump and hoisted myself up. Crack! as my head met a low rafter. There was a scuffling below as Elisabeth dodged out of the way.

"You all right?" she asked.

"Just hit my head," I moaned.

"Yes," she replied, "the roof slopes down there."

I sat down on a joist to rub my head and consider matters. Beside me stood a huge metal-lined box.

"Hey, what's this big box?" I shouted.

"The water tank. The pump fills that, and the water comes down from there to the toilet and the sink."

"Gravity feed, eh?" I observed in my best plumber's voice.

"Richard! Get down off that ladder."

"Want to see what Daddy's doing," he complained amid sounds of a struggle.

"Hey, run the pump," I said. "I want to see what happens." She thumped downstairs to the kitchen and, after two or three reluctant grunts, the house was motorized. The joist on which I sat buzzed in sympathetic vibration with the engine in the kitchen two floors below, and the attic throbbed with the sound of muffled machine guns. After a few seconds an inch-wide stream of cold water ripped into the tank. I bathed my hand in the icy jet. Marvelous, I thought. The river, maybe fifty feet below me, and that racket in the kitchen brought it pouring into this tank in our attic.

My legs dangling through the trap door, I mused on the miracles of science. The thin stem of Elisabeth's voice rose from the chaos below.

"Give me a shout when the tank's full and I'll shut it off." It was a big tank. I hadn't meant to sit there admiring the genius of science that long, but there was too much noise for debate. After twenty minutes the water had climbed to within an inch of the top.

"Okay!" I called down through emptiness. No answer. *"Okay! Shut it off!"* I shouted. Still no reply. The water was rising. Reaching out and finding the joist on the other side of vacancy, I swung myself into the void. I pawed with my feet for the top step. It wasn't there. I would have to drop lower . . . no, even lower. It must be there . . . somewhere. She couldn't have moved the ladder. As I bent my aching arms, I wondered if this was what all those chinning exercises in school were to prepare me for. The ceiling was just too high. I would have to let myself go, hang down completely before I could reach the damn ladder. The tank must be brimming by now. I dropped and swung for one eternal instant before kicking the ladder. As I reached the floor, an icy torrent poured down my back. I ran down to the kitchen and shoved in the choke on the thundering engine. As it died, the sound of our private Niagara filled the house.

"Buckets, dear," cried Auntie Em, "get the buckets, dear. It's happened again." She and Elisabeth mopped furiously until there was only the peaceful dripping down the stairs, as from the eaves after a heavy rain.

"I kept calling you," I said.

"I was just making the beds," Elisabeth said angrily. "I couldn't hear you because of that *engine."*

"How about a nice cup of tea?" suggested Auntie Em. We took it out onto the front porch, where we dried off and cooled down.

"All blue and silver through the willow tree," Elisabeth mused, half to herself. The voices of Richie and Mark drifted up from their game on the shore. "They love it too," she said thoughtfully.

"What about your candle doings, Bob?" inquired Auntie Em, breaking the spell.

"All soaked, I suppose," I replied.

"I see. I see. So what precisely do you think we ought to do?" she pressed. "Shall we have that bug business . . . that spraying?"

"At the moment," I said after a pause, "at this very moment, I think I'd like another cup of tea." "Damned escapist," part of me accused; "face up. The bats won't go away by themselves." Reluctantly I faced up, but the problem seemed just as hard. "Probably not such a great idea to pump the attic full of insecticide. Water supply is up there."

"Then what are we going to do," Elisabeth urged with merciless persistence.

"How about stuffing the bits of paper and rags back in?" I offered.

"Quite, quite," Auntie Em chuckled.

"Tell you what"—the coward in me rising—"let's have a swim and think about it." So we did.

Next morning, full of courage and armed with a hammer, a pocketful of nails, and a bundle of long strips of light wood, I mounted the wobbly stepladder again to tack the strips around the edges of the ceiling upstairs, thus—I hoped—closing off most of the cracks through which the bats came to visit. I had hammered a few nails before, but not very many over my head, and none that I could recall from the top of an arthritic stepladder. Every blow brought a fresh shower of bat droppings down upon me. The only thing going for me was that I didn't have to look up to see what I was doing. Elisabeth's interruptions didn't help either.

"I wish you'd sharpen that scythe like you said you were going to." . . . "You know what that bloody Richard just did? He pushed Markie into one of those thorn trees. They're a real hazard. We've just got to do something about them." . . . "Bob, the toilet is running and running. I can't stop it. You'll just have to come."

"Call the plumber," I shouted. "I don't know anything about toilets! Did you learn anything about toilets at Oxford?"

"What's that got to do with it?"

"Well, they didn't teach me anything either," I roared, giving my swollen thumb another sound whack with the hammer. Swearing and groaning, I came down and went into the bathroom. The mysteries concealed by the lid at the back of a toilet had never excited the mildest curiosity in me, but I had no choice, plumbers being where they were and we being where we were. I lifted the lid of this one and reached in and met, for the first time, that strange, cranelike mechanism, the intricacies of which I was to puzzle over many times in later days. There was a sort of rubber ball stuck on the end of a vertical rod. The rod had become stuck in its guides, and I inadvertently unstuck it so that the ball dropped down, sealing the outlet from the tank into the bowl. Miraculously, the tank began to fill.

"Jolly good!" cried Elisabeth. "Well done! I say! Isn't that great? You've fixed it." Spurred by praise, I examined the "doings"—Auntie Em's word for anything too complicated to explain or understand—to find out exactly what I had done. Pulling up the rod, I lifted the ball from its seat and the water rushed again into the bowl.

"Whoops! What happened?" Elisabeth exclaimed.

"It's all right," I reassured, dropping the ball back in place. "Just testing." And then I lectured her on how toilets work.

"But why," she asked when I had finished, "why does the water stop coming into the tank when the tank is full? I mean, why doesn't the tank just keep overflowing? How does it know when it's full?"

"One of those miracles of science we keep hearing about," I declared firmly, but it bothered me too. Sweeping it aside, I climbed back to my nailing.

Those strips actually did help. They reduced the nocturnal invasions to sorties from odd scouting parties. Having made some progress, I tried to figure out how to make more. As I lay awake puzzling it over, I could hear the gentle ghostly chime of empty coat hangers in closets announcing that another visitor had landed and was swinging slowly about and considering his next flight. Seeing no solution, I finally admitted one morning over breakfast on the back porch, "Sometimes you've just got to compromise, that's all. Let them have the attic. What do we want of the attic anyway?"

"Water, that's all," commented Elisabeth. "Richard, take your plate in, if you're finished."

"How about moving the jolly old water tank downstairs, Bob?" asked Auntie Em.

"Can't really," I answered.

"But you're so handy with those hammer things," she encouraged. "I expect you could do it if you really put your mind to it."

"Maybe I could move the tank, Auntie Em," I admitted, "but I really couldn't do much about gravity."

"You don't know your own strength," she pursued.

"But, you see, Auntie Em, the tank has to be up above the bathroom for the water to run down from the tank into the sink and the toilet."

"I see, I see. But I expect you could do it all the same, what with all your doings and so on." After a pause she continued, "My gracious! It is so nice to have a man about the place."

"But . . ." I objected.

"By the way," Elisabeth observed, picking up the breakfast dishes, "I think we're about out of water. How about running the pump?"

Going into the kitchen, I gave the starting cord a few yanks and it thundered into action. I returned to help with the cleaning up.

"I say, Bob, hadn't you better check up there?" inquired Auntie Em nervously.

"Lots of time," I said. "Takes twenty minutes to fill that tank."

"I see," she said doubtfully, "but you can't be sure how much is in the tank now, can you?"

"Lots of time," I reassured, but she wasn't. She made that clear by the way she bustled about with the dishes and clanged the kettle on the stove. Every clatter she made fairly shouted: "You *aren't* sure, and you'd better get up there." As her clamor mounted, I became more determinedly casual. Humming abstractedly, I hunted around in the shed for nothing in particular till the tension finally got too much for me, and I walked slowly through the roaring kitchen to where Auntie Em was sweeping the dining room with furious hisses of her broom.

"Here it comes!" shouted Elisabeth, tearing past me to shut off the pump. As the engine died, the rushing of cataracts filled the house. Auntie Em's stiff British upper lip was so stiff that, clamping down, not one single denunciation passed forth as she dropped her broom and ran for a mop.

"You!" Elisabeth hissed at me with a suppressed snarl as she mopped.

"If you're not careful, I'll put those rags back in the wall," I whispered.

As the last few drips splashed peacefully into the closet under the stairs, Auntie Em had recovered herself and chirped, "Well, we do have our times, don't we?" After a pause she added, "But you will think about it, won't you, Bob?"

"About what?" I asked suspiciously.

"About moving that jolly old tank downstairs. It would be trying, I expect, but once done, it would be right where anybody could find out just how full it was. Perfect place, that closet under the stairs, for example. A person would just have to poke in his head to see what was going on with the doings."

"Yes," I said sadly.

"Good-o!" she cried, "I knew you could settle things. Just a few of your hammerings, eh, and then we shall see what we shall see! What, what?"

"She's your aunt," I muttered to Elisabeth that night in the safety of our bedroom. "You explain gravity to her."

"Hush! She isn't deaf, you know; and anyway, she's my *great*-aunt. Besides, why can't you move it downstairs? Sounds like a fine idea."

"The tank won't fit through the trap door," I explained wearily.

"Ah! Why didn't you say so in the first place?"

"So that's it," exclaimed Auntie Em in the morning. "The doings won't fit through, eh? I knew you would . . . figure things out, Bob. Gracious, how I wish I had you in my little house in Cornwall! You'd have things straightened, you would." She considered things. "But, Bob, how do you suppose they got the jolly old thing up there in the first place? How did they?"

"Took some doing," I answered. "Probably built it up there."

"Built it up there—that's what they did. They must have done, if it won't fit through. In Ceylon they were always doing that sort of thing. [Three decades earlier she had spent a year there.] They were always doing that sort of thing."

"And what did you do in Ceylon, Auntie Em?" I asked, hoping to get her off the water tank.

"Took care of the children," she explained. "A sort of school it was, do you see? Little beggars, they were, just like Richard here. Never could tell what they were going to do next. How they did keep me hopping! But they had elephants for water. Barrels and barrels they'd carry. What you need is some elephants, Bob!" and she laughed. "Ah, we did have our times, we did. It was beautiful out there too, but not as beautiful as this. I expect this is the most beautiful place in the world. Dear me, yes! Ceylon was beautiful, but one can't compare it with this, with that willow tree down there by the water."

"Even with the bats?" I put in. "We've still got bats."

"Oh yes, yes we do," she agreed, "but in Ceylon, Bob, why they had snakes, great big rotters. Oh no, give me bats every time. Besides, there aren't so many of them since you've done your hammerings and so on. Not nearly so many. Didn't I tell you that Bob would take care of things when he came? Didn't I, dear? What, what? Then why don't you two go out for a row or a fish? I'll keep an eye on Richie and Markie. We can go for a walk. I can still take care of children, you know."

"Not a bad old stick, is she?" Elisabeth said as we went down to the boat.

But usually when we walked down to the boat it was not with the happy prospect of a peaceful hour on the water.

On the mainland, buzzing down to the supermarket is only a nuisance. Preparations there consist of checking for car keys and money. It isn't like that for an islander. It takes planning, because it isn't the sort of thing one wants to do again right away—at least, we didn't.

For fifteen minutes Elisabeth would sit hunched over the kitchen table, pencil firmly clenched over a bit of paper, teeth set with concentration, mumbling to herself. Finally: "Anyone think of anything else?" she called.

"What have you got on the list?" I asked.

"It's too long to read. Everything."

"Then what's there for us to think of?" I pointed out. She stamped her foot angrily.

"Butter, dear," called Auntie Em. "We were always forgetting the butter in Ceylon."

"I've got butter," Elisabeth called back.

"Hinges for the screen door?" I asked.

"Which kind?"

"Ice, dear," Auntie Em chimed; "we're just about out of ice."

"Oh Lord! *Ice!*" She scribbled.

"And milk," added Auntie Em.

"Got that."

"Spring," I said.

"Spring? Spring what?" Elisabeth asked.

"Spring *hinges,*" I said.

"Are there different sorts?"

"Of what?" I inquired. "Different sorts of spring hinges? Sure there are different sorts."

"Mom, can I come?" Richie piped up.

"You'll have to come anyway, Bob, to carry the ice, so you can pick your own hinges."

"Markie come too," he urged, pulling at her dress. "Markie come!"

"And white gas for the lantern," I added.

"Beastly heavy, that ice!" Auntie Em chirped. "How it used to melt! How those men used to complain about having to go for more! Oh no, they didn't like it, not one bit."

"Tea!" Elisabeth exploded. "Almost forgot the tea. Anything else? Listen, everybody! Can anybody think of anything else? Because I'm not going again this week, hear?"

"Can I come?" persisted Richie.

"Markie too!"

"Don't forget the garbage," she declared firmly, rising. "Don't *any*body forget the garbage." There were big refuse bins by the town dock for the use of islanders, and we brought in what couldn't be burned. "She'll have to come too, Bob," Elisabeth said quietly. "She loves to go to town. Wash your face, then, Richard. You too, Markie, and get some shoes on."

"That many in the boat, it isn't safe," I objected softly.

"But she can't just sit here all the time." Turning, "Would you like to come, Auntie Em?"

"Oh, I say! That would be nice. Better get smartened up, though," and she went upstairs briskly.

"Well, we're not taking that damn dog," I declared angrily. She made no answer. Flory was our half spaniel, half labrador who didn't like boats, but he downright hated staying home alone.

With Elisabeth, the children, and the garbage safely in the boat, I helped Auntie Em down as I alternately ordered and threatened Flory to stay back where he was. Auntie Em safely seated, I shouted one last stern command, dropped in, and gave us a quick shove away from the dock. Flory dashed to the spot where the boat had been moored and measured the distance for a leap. With a howl of dismay, he decided against it. But he wasn't ready to give up. Barking and carrying on, he ran back to shore where he bounded along, determined to follow us as long as he could. With stony indifference I heard his yelps and howls over the spluttering motor. I didn't relent, but then, Elisabeth was steering the boat.

"Poor chap," she sympathized. "Poor chap! He's so miserable, and he might get lost," she said, nosing the boat into the next dock.

"Lost?" I cried. "On an *island?* Where the hell can he go?" But it was no use, and Flory sprang in, tipping us badly. Eagerly climbing around and over Richie and Mark, he put his paws on my knees and licked my hand.

"Don't thank me," I growled. "It's her fault."

"Remember, dear," Auntie Em declared with firm solemnity before the motor consented to go again, "remember, I don't swim. But if anything happens, I've had a good inning, so don't worry about me. Save the children." (We remembered. It's still our watchword.)

"What's going to happen?" Richie inquired nervously.

"Nothing," I replied sharply. "Nothing's going to happen. Just take it easy." Over the sporadic putt-putting of the motor, from far up the channel behind us, came the ominous drone of a tour boat, the Thousand Island equivalent of a sightseeing bus. On the mainland, being passed by a sightseeing bus is scarcely worth mentioning. In a twelve-foot car-top boat carrying five people, a restless dog, a garbage can, and an extra tank of gas, being passed by a huge tour boat under full power is a fearsome event. Their wakes have been known to snap the lines of a moored boat and heave it up onto the dock.

"Tour boat," I commented casually.

"The doubledecker," Elisabeth observed through pursed lips as she turned to get as far out of the channel as she dared without running aground. The drone grew to a roar. Over its thunder the loudspeaker placidly announced to the passengers: "On our left is Polaris Island, owned by Captain A. Winfield Chapin. The cruiser you see tied to the dock is the two-hundred-fifty-thousand-dollar *Captain's Gig.* If you look carefully, you will see the mailbox at the end of the dock. Residents of the Admiralty Islands receive mail daily by boat from the mainland."

I hoped we could all be back to receive our mail by boat from the mainland next day. Elisabeth swung our bow to meet the wake.

"Hang on," she said tensely, "here it comes!"

"Oh dear," cried Auntie Em as the first wave lifted us. "Oh *dear,"* as the second raised us higher and we sank into the trough. *"Oh dear!"* as the third hoisted us high and the crest broke over the bow and we shipped a few pails of water. And yet again, as the fourth seemed to point the bow to heaven and more of the St. Lawrence poured in. Then the wake was passed.

"Now it's bailing time, Rich," I sighed with relief.

Branded in our memories are those times when, after making a safe landing, we drove up to the main street. Then Elisabeth's groan of dismay. "It's *closed!* My God! They're all *closed! Everything's closed!* Must be Wednesday afternoon!" Except for the restaurants and tour-boat line, Gananoque shut down every Wednesday at noon.

"Could be Sunday," I suggested. Without electricity, we had no radio, no disc jockey announcing the time every five minutes; and after the first week we forgot to wind our watches because, after all, what difference did it make whether it was seven-thirty or ten o'clock? When we were hungry, we ate; when it got dark, we went to bed, and when the birds woke us we got up. Whether it was Monday or Thursday made no difference. It did matter, however, on Wednesday afternoon or Sunday. We weren't unlucky often enough to make us keep track of the days. Being freed of clocks and calendars came as an easing of the spirit.

Even when the town was open, it was hardly a picnic. After two or three hours sweltering up and down the main street, we loaded the boat, wondering whether the fifty pounds of ice was actually going to sink us, whether the motor was either going to start or give out, and whether we would meet the tour boat again—and this time we would be fully laden. The motor did often die; sometimes we did meet the tour boat; sometimes even both happened at once, but we never quite altogether sank. Frequently our good neighbor, Dick Johnson, met us at his dock to help carry the groceries up to the house with his wheelbarrow.

"It is a business, isn't it, this going to town?" he would say. We agreed. But once everything was stowed away, relaxing on the front porch with cups of hot tea and heavy chunks of "olde English fruitcake" reduced for quick sale from the A&P, the wind blowing, sun shining, and the river rolling blue and silver, the catbirds and song sparrows singing, we settled gently into the peace and joy of being islanders.

"Isn't it just awful in town?" Elisabeth exclaimed. "So *hot!* I wouldn't be on the mainland for anything."

"Dear me, no!" agreed Auntie Em. "Such *doings* they have over there! *Such* doings! Now, Ceylon was an island—I expect you know that, Bob—but it wasn't like Hay Island. Dear me, no. Why, it was so big that it might just as well not have been an island at all. No, no. This is the place. Such times we do have. Such times! Too bad Joanna couldn't be here." (Joanna, Elisabeth's sister, had come to the States the previous September and had been teaching in a girls' school in Baltimore.) Elisabeth tried gently to break her news about Joey to Auntie Em.

"She's very busy, you know . . . very, very busy and enjoying herself. She's met a . . . a very . . . a very nice fellow."

"Has she, though? A young man, eh? Well, now, isn't that nice. She always had plenty of get-up-and-go."

"Yes," continued Elisabeth, "and she seems to like him . . . like him very much." After an ominous pause, Auntie Em said suspiciously, "Well . . . I suppose that's quite nice for her. Yes . . . yes. I suppose it is."

"So much," Elisabeth pursued doggedly, "so much that . . . she may even be thinking of getting married." A stunned silence.

"Not . . . not really, I don't suppose," Auntie Em reassured herself aloud. "Not actually married! You don't mean *married!*"

"Wouldn't be surprised," commented Elisabeth. "In fact, I think so."

"Oh dear . . . *oh* dear . . . dearie *me* . . . really *married?*" Auntie Em's seventy-five years of maidenhood were having trouble swallowing so bitter a pill for her beloved great-niece.

"Bob and I are married," Elisabeth encouraged.

"I know, but . . . but . . ." Her rocker creaked rapidly back and forth. "The shock . . . the shock. I believe I'll just lie down for a bit." Elisabeth helped her to her feet. "Such a lovely girl too," she mourned with the first touch of feebleness I had ever heard in her voice as they went inside. Slowly they mounted to the second floor. "She's actually taking to her bed," I thought in disbelief. This model of British pluck, this cheery, bright, sparky . . .

"Going to get some brandy," Elisabeth cried, bursting past me where I still sat on the top step.

"We've got home brew," I called after her.

"My God!" is all she threw back over her shoulder as she ran to the dock. And the motor coughed as she set off toward town on her errand of alcoholic mercy.

A couple of hours later, propped up in bed and sipping a thimbleful of blackberry brandy, Auntie Em forced out: "Can you . . . can you tell me anything . . . anything about this . . . blighter?"

"He's not a blighter, Auntie Em," I said, choking back a laugh. "He's a very nice chap."

"You know him, then?"

"Sure. Of course."

"And is he . . . can you tell me whether he is a . . . a Christian? Is he?"

"Do you mean, does he go to church?" I inquired.

"I mean," she said sharply, "I mean, is he a *Christian?*"

"Yes, he's a Christian," Elisabeth put in. "And he's a very nice chap. He'll be very good to her." Auntie Em sniffed her disbelief.

"But," she declared, rising up in bed as if to scotch this blighter utterly, "but is he *British?*" Tense, she waited for the answer.

"Well, no, he's not British," admitted Elisabeth reluctantly. "He's an American."

"Ah! I thought so! I *thought* so," and she subsided, his perfidy forever settled as far as she was concerned.

"But that's not the end of the world," Elisabeth soothed. Auntie Em was resolutely silent. "I mean, Bob is a Christian . . . well, American, after all."

"Yes," she observed dryly. *"And* a *man,"* she muttered bitterly. After a pause, "Poor child . . . poor child!" she mourned softly.

"Who? Do you mean Joey?" asked Elisabeth. "She's really not a child, you know. She must be twenty-seven."

"She could have done so much with her life, so much. A lovely girl."

"Not a girl, really, Auntie Em," Elisabeth urged gently. "She's a woman. Really she is." Then softly, "Bob, why don't you go check on the children?"

I accepted my dismissal with relief. I went downstairs quietly, made certain that Richie and Mark had their life jackets on, opened a bottle of our foul-tasting but potent home brew, and settled down again on the top step to listen to the river.

American, and worse still, a male. I wondered at the peculiarities of Britishness and seventy-odd years of maidenhood. My wonderings were not without a cutting edge of amusement. Secure in my own convictions and point of view, I sat contented in the lap of summer, enjoying her caresses until, inexplicably, the song of a catbird turned my thoughts backward upon myself. To him . . . what must I and my carings seem like if looked at from his swaying branch? An awkward, lumbering, earthbound ox—without wings, without song—unable to weave even the simplest of nests. I shivered. But need I go that far? What about my own kind—those others, my fellow creatures, back on the mainland living their normal lives—what would they think of the whole pack of us if they knew?

Well, I decided, we had all better have the right to appear a little absurd in somebody or other's eyes. That right denied, everybody would be in pretty bad shape.

When Auntie Em took ship for home, she left her special stamp on that summer. She had shared with us her pleasures, her fears, and, most of all, her self. Through that sharing, she had become part of our adventure and, therefore, part of us.

3

Pike Fishing by Bell

Every river, pond, or stream a fisherman passes is an invitation and a challenge. But, despite the seductiveness of all water, each has his own special spot, the place where he feels most away from things, the place he feels most himself.

For me, that special place became the water just off our shoal. After a scramble over the loose boulders to where it lay, I perched on the storm-polished tree trunk, rod in hand, the tangled loves of family in the house behind me, the unknown sliding by in front—sitting, as it were, on the very edge of consciousness and tossing out a lure in the hope of making contact with a something that might be moving about somewhere past that edge, that boundary where the me stopped and otherness began. Sitting there, slowly the tensions of connected thought relaxed, sagged, and finally fell away. Hair-line cracks began to appear in the corners of my mind, eased open, splayed out, and let in the quiet. Once inside, it rose with the gentle insistence of a tide, until it floated conclusions loose from their premises, dreams from their moorings of hopes. At length the tiny iridescent globe of self drifted on its surface as gently as any fisherman's bobber.

When one has achieved this state of suspension that every fisherman knows, the conditions are right for revelation. He is alone, and from his aloneness he casts his line out and through the surfaces of things to ask if another lives. With infinite expectation he awaits a word. Suddenly along that line comes the electric news—the unmistakable telegram declaring that he is not alone. The fish takes the bait, shakes it, lets go, grabs again—more securely this time—then runs with it to a place of safety where he mouths, tastes, chews, swallows, and, his hunger satisfied, moves off sedately. All this he shoots along to the fingers trembling with excitement. In a random world, that tenuous thread one has flung out in hope has miraculously been enough—contact between self and self suddenly exists. It is epiphany—a fact. As He was a "fisher of men," so are we all. And how many lines do we cast into that which is beyond? How rare, but how miraculous the bites! And how often do we ourselves take the bait another has cast?

So, sitting on my timber, fishing . . . of themselves images rise and move slowly across the screen of memory . . . images of that first trip when, for me, the world was still dressed in that glory which sight confers, that special sheen only the eyes of a child can see. Perhaps I was five when my brother Bud, who was twenty, took me with him for a morning's trout fishing on a stream near our home in southern New York. When we reached the creek, he plucked and trimmed a small branch. Taking a four-foot piece of line, he tied it to one end, and, making a noose at the other, slipped in a worm, drew the knot tight, and handed me the branch.

"What do I do with this?" I asked.

"You fish," he explained, and started along the bank with his rod. I shouldered my first outfit and trudged behind him. For a while I imitated his every move, but then I saw a bright green creature leap from a log saying, "Kgunk!" before he disappeared.

"Hey Bud! What was that?"

"A frog."

"Are there any more?"

"Sure. Lots, but you've got to look for them."

I started looking. I didn't see another frog, but, under a clump of grass, I found a dark little hole.

"Hey Bud, Bud! What's *this?*" I called. He turned his head.

"I don't know . . . maybe a muskrat hole."

"Is he in there?"

"Call him and see," he threw back.

I called, but there was no reply; so I took a stone and knocked on the earth beside his door and called again. "Hey! Are you in there?" Still no answer.

"Come on," Bud urged. Reluctantly I rose and started down the bank. Accidentally I kicked over a large stone, and out from under it scuttled the strangest creature. He was covered with a shell the color of peanut brittle, had about six legs and a big flat tail that he kept sort of scooping himself backward with. And as he scrambled along he waved these two big claws in front of what ought to have been his nose and kept making fists with them.

"Hey! Hey *Bud!*" I cried. "What's this?"

"What's what?" he asked a little irritably. "That? That's just a crawdad."

"Can I have him?"

"What do you want with *him?*"

"I just want him, that's all."

"I guess so. Wait, though. Don't touch him. You've got to pick them up in a special way or they'll nip you." He stopped and picked up the crawdad by the back between his thumb and forefinger. "Get me that old coffee can over there with some water."

I picked up the rusty can, dipped it half full, and held it beneath the crawdad. Bud dropped him in. I spent the rest of the morning peering into the can to watch his quivering antennae and the way he could scrunch himself along backwards, all huddled up in an armored ball with those big claws clip-clipping out in front.

We came home without any fish. But I had come home a fisherman, and for life.

So, twenty-five years later I sat at the end of our shoal casting into the mighty St. Lawrence—my Paradise. But even in Paradise there was a snake, and my special place had its drawbacks. Dropping off as it did into sixteen feet of water, my predecessors had looked upon this spot as a perfect dumping ground for old bedsprings, the broken porch glider, and discarded kerosene stoves; so I lost more than my share of lures there. When I hooked a good one, he would dive down into this jungle, take a couple of loops around a jagged bit of iron, give a couple of yanks, and swim calmly off about his business. Yes, even that spot had its drawbacks. I longed for a companion with whom to explore and fish other parts of the river, but neither Dick Johnson nor his father were fishermen. Elisabeth would always be eager for an expedition, but there were children. Of course, I had a father's responsibility, which includes that of initiating his offspring into the joys of fishing; so, as often as we could, we did try family fishing excursions.

Snapping Richard and Mark into life jackets, Elisabeth and I put them into the boat. Before starting the motor, I delivered my stern lecture on "sitting still." Then, after a couple of pulls, the motor consented to serve. Elisabeth slid into position in the stern and I moved to the middle seat. We were off!

Before clearing the dock, Richie started flailing an old casting rod, the tip buzzing past my ear. I cried, "Hey! Richie! What are you doing?"

"Fishing!" he shouted back.

"Give me that rod!"

"Bob!" called Elisabeth, "make Mark sit down. He's leaning over the side at the bow."

With Richie's casting rod in my hand, I worked my way forward, kicking over the bought worms on the way, to where Mark was hanging over the side trailing his hand in the water. Yanking him back with my free hand, I put down the rod and smacked him. Over the sound of the motor and his howls, I launched again into my lecture about "sitting still."

Whack! The tip of my spinning rod, with which Richie was now "fishing," hit the back of my neck.

"This is good enough," I shouted to Elisabeth. "Cut the motor."

"What?"

"Cut . . . the . . . motor!"

When the engine died, Elisabeth said. "What's the matter with the motor?"

"Nothing. Nothing. This is fine. We'll try it here. Richie, pick up the worms."

"Daddy," whined Mark, "my line is tangled. Can you fix it?" I started to work on the rat's nest he handed me.

"Will you bait my hook?" called Richie from the bottom of the boat.

These were short trips—not at all the sort of thing that comes to mind when one thinks of a quiet afternoon's fishing.

I had to figure out some way of getting out on the river alone.

The answer came from an unexpected quarter and on a very unlikely occasion—graduation day at the College. The faculty were lining up in the hall of the Liberal Arts building in full academic regalia for the procession down to the football field to swelter through another of those grim ceremonies. Durrell Enck, the chairman of the Physics Department, whom I knew only slightly, came up and said. "Hi, Bob. How about being my partner in the procession? English and Physics make a good combination."

"Thanks, Durrell," I answered as he took my arm and we started out into the broiling sun. As we entered the field and made our way toward the seats of honor, I said, "I hope you don't have to stay in Lancaster all summer. If you do, you'll be done to a crisp by fall."

"Yes, I'm going to be here. I'm teaching summer school. What about you?"

"Not me. I'd rather starve on my island in the St. Lawrence." We reached our seats and sat down.

"You lucky dog!" he murmured. "I bet it's nice up there."

We rose for the invocation. When it finally ended, we subsided.

"It's marvelous," I whispered. "Always a breeze, and you can go for a swim any time."

"And I have the signal honor of introducing our speaker on this great occasion . . ." droned the President.

"Any fishing?" Durrell asked.

"Pike and bass. Mostly pike, northerns."

". . . and all you young men with shining faces on the threshold of life . . ." Jolly funny place for their faces, I thought.

"Get any big ones?"

"Sure. Why don't you come up and try your luck?"

"I'd like to."

". . . and from that future which now stretches before you, you will look back to these days, your eyes shining with the tears of memory, and you will say . . ."

"Come up after summer school. Bring your family. We've got lots of room."

"Serious?"

"Of course. Just drop me a card saying when you'll arrive, and we'll meet you at the town dock. Then I can show you my problem."

"What's your problem?"

". . . these learned doctors, these sage and distinguished men who sit in proper state before me . . ."

"Well, it's rather involved."

"Never mind that. Shoot. I'm listening."

I explained about the bedsprings and old stoves at the bottom of the river and what it was like trying to fish four people in a boat when two of them are under seven years old.

"Somehow I've got to get out on the river by myself. I need a way of keeping in touch with my dock—a radio transmitter or something like that." He was silent for a couple of minutes while the speaker ground along.

"Have you got electricity?"

"Yes, just got it."

"Well, what about a bell, or something like that?"

"That's an idea, but I can't have a bell ringing solidly while I'm out fishing. I would be going out around five in the morning, and I have a neighbor nearby."

"Suppose it rang just once in a while—maybe every twenty seconds?"

"That would be good. That would really do the trick."

Everybody was applauding because the speaker had finished. The President started conferring honorary degrees on visiting dignitaries and the Dean began handing out prizes.

"Do you think you could work out something like that?"

"Tell you what. Why don't you come over to the lab late this afternoon and we'll see what we can rig up. I can't promise anything, but we'll have a shot."

At four-thirty I found Durrell in the lab working with a small electric motor the size of a fat pocket watch. From its center protruded a shaft that it turned at the rate of one revolution a minute. To this shaft he had attached an ordinary radio knob, but, instead of leaving it circular, he had cut the sides into a square. He was mounting the motor with its revolving square knob in a little wooden box.

"What I'm going to do," he said, "is put a set of flexible metal contacts right beside the motor so that each point of the square knob will press the contacts together for a second. They'll spring back, breaking the circuit when the point passes. When the contacts are touching, they will let the current go through to this big old bell, and it will ring."

"Will it work?" I asked.

"It will before I get through," he said; and it did.

The next day I took Durrell's invention up to Hay Island. I mounted the bell on a post at the end of the dock, ran two hundred feet of lampcord up to the cottage where I had installed his box according to instructions, hooked it up, pushed in the plug, and started for the dock. Halfway there, I heard the chime of the bell.

I found the first mooring ring, cast off the stern line, and stepped into the twelve-foot dinghy. Pulling the boat along the dock with my hands, I came to the ring to which the bowline was tied. I cast it off, sat down on the middle thwart, and pushed free of the dock. I glided out to the "ding-ding" of the bell. Bare feet thudded along the dock and Richie's voice called. "Hey Dad! where are you going?"

"Just out for a little row." I pulled gently on the oars and the boat slid out toward the channel.

"Can you go out by yourself, Dad?"

Ding . . . Ding.

"Sure."

"Will you take me for a little ride?"

"Have you got your life jacket on?"

"Yep."

"Okay. Hang on." I backwatered till the stern bumped gently against the dock. He climbed in. We had our ride and I docked again easily.

The gate out onto the river had swung open for me.

The next morning I slipped out of bed at five, dressed quietly, crept downstairs, took my box of lures and spinning outfit, and started for the river. I was in the boat before I remembered I hadn't plugged in the cord. This done, I picked up the oars and pushed off onto the velvet cushion of still water away from the sleep-filled house. Every fifteen seconds, as each corner of that square knob closed the contacts, the bell spoke—telling me all I needed to know. I pulled strongly for the southern tip of the island.

The catbirds, yellow warblers, song sparrows, and whitethroats were waking, and as they tuned up they sang that the shore was about fifty yards to my right. I rowed softly on through the silent water along the line of their singing, keeping the sound of my bell astern and slightly to the right. The gulls began to cry and complain over the river. Far behind me and to the left from somewhere out of the Forty Acre Shoals came the fluting of a loon. Way out to my left, perhaps a mile, an outboard started up and whined off toward Gananoque. Minutes later, its wash passed and slapped up on the rocks, confirming my position. The sun had risen high enough to send a warm beam to my right cheek, and a light southwesterly breeze sprang up at my back.

The bell, which had unchained me from my dock, was now far astern, and its quiet "ding . . . ding" had almost blended into the other morning sounds. I pulled hard on my right oar, sending the bow farther out into the channel. I coasted for a moment, enjoying my freedom and the morning. Then, reaching for the anchor line, I released it, and the little pulley squeaked and rumbled as twenty feet ran over it. The anchor down, I let out ten feet of slack, secured it, and lit my pipe.

I snapped an abu onto my swivel and started casting.

With each toss I let the lure sink to the bottom and then brought it back in short two- or three-foot runs. On the third cast I got my strike, a firm, solid tug. I set the hook and the reel buzzed. I hadn't even moved him. Slowly I cranked the reel and the weight at the other end came gradually up and in. Then the rod suddenly bent and the reel screamed as the fish, realizing he had been hooked, dove for the bottom. A pike, then, and a big one!

When he stopped his dive, I started gently pumping the rod. With a big pike on four-pound test, I couldn't afford to get horsy, and I wanted this one.

Slowly he gave ground, but when he was up four or five feet he plunged again. The rod doubled and the reel buzzed. This time he wasn't going to move. For three minutes he lay there stubbornly resisting. Then, suddenly, he rushed toward the surface. I cranked furiously to keep up with him. The line became taut again with that same loggy heaviness. Slowly I led him toward the boat, and he came. When he was about five feet from the tip of my rod he seemed to catch sight of me and, with a great wallow and thrash, he set my reel shrieking as he sounded again. I let him go and, when he had finished, began urging him gently up. I couldn't rush him, not only because of my light line but also because I wanted him really played out for landing.

You can't net a fish you can't see. Landing a perch, a bass, or even an eel is easy. I play them out, then pass the butt of the rod along till I come to the tip, then feel down the line until I reach the fish, grab him by the gills, and pick him up. But pike have teeth, big ones, like needles. Their mouths are large enough to take a man's hand comfortably.

This one I had hooked was a big one, so I wanted him dog-tired by the time I began reaching around for his gills.

Gradually he came up, then dove again. He repeated this four times. Finally he was through fighting and seemed to lie near the surface, daring me to finish the job. By raising the end of my rod I could tell he had only three or four feet of line. It was time.

Easily, so as not to upset him, I passed the butt of the rod to the stern till I reached the tip. Holding the hair-thin monofilament between the thumb and index finger of my right hand, I worked my left down the line about a foot. He would explode if I touched him, so I moved the line from side to side, trying to judge how far away he was. He drifted toward me. I thought I had about another foot. I let go my right hand and pinched the line just above my left, then moved my left down another six inches. He was close now. I was leaning far over the gunwale. I moved my right hand down to my left, pinched the thread again, and let go with my left.

Moving the line back and forth and raising it a couple of inches, I knew he was very near. There was no give to the line, and the arc in which I had to move it was very sharp. I would have to make my grab—now! before he started thrashing. But which way was he lying? With his tail toward the bow or stern? I had to gamble.

I moved my left hand down the gunwale about eighteen inches toward the stern, reached out as far as I dared, and, fingers spread, plunged my hand into the water as deep as I could and scooped back to the boat. He was beneath it. My fingers slipped along the underside of his lower jaw. I had missed, but so had he.

Instantly I pulled about a foot of line with my right hand and made another grab. My fingers closed on his throat just behind his gills. I pressed him against the side of the boat and slipped my fingers into the gill slots. When I had a firm grip, I hoisted him out and into the boat, which righted itself.

When he hit the boards, he spat the lure and began to thrash. Holding him with my left hand, I reached under my seat with the other till I found the sawed-off stub of an oar. Holding his head slightly off the bottom, I gave him three solid whacks on the forehead. His great body quivered and was still.

I made sure he was through before I began hunting around in my pockets for my Braille tape measure. Thirty-two inches, he was, and, by the feel of him, he would go about eight pounds—eight and a half, as it turned out.

I dropped him back onto the floor boards and straightened up. It took me four matches to get my pipe going again. Then I sat there—calming down and slowly rejoining the rest of the world.

Far behind me I heard the day's first tour boat. It was coming right down the channel. Its engines would drown out the sound of my bell, so I would have to wait until it had passed before starting for home.

The tour boat swung out to my left to avoid swamping me with its wash. When it drew abreast, I reached down, picked up my pike, and held it high over my head.

The thunder of the engines died and the big boat idled over until it was almost alongside me. I held up the pike for as long as I thought they might be interested, then laid him down.

"Hey," called a voice from the deck far above me, "hold him up again, will you? I didn't get my picture yet."

I held him up.

"Thanks," he shouted.

As the big boat turned away, the purr grew to a growl and then a roar, and away went the crowd of tourists, each of them having something I should never have—a sight of my fish, my green-and white-bellied fish with the long savage jaw. But then, I had the fish.

As the thunder of the big boat faded, it left the river and the morning to me and to the gulls and to the whitethroats and redwings on shore, and in the distance astern came the soft, regular "ding . . . ding" of my bell. I drew up my anchor, secured the line, and picked up my oars. As I rowed back along that thin, clear thread of sound, I had the wetness of the great St. Lawrence on my palms, its sounds in my ears, its smells in my nostrils, its joy on my tongue, and, at my feet, an ingot from its dark and inexhaustible treasure house.

4

The Fiberglass

"It's very kind of you, Mr. Johnson," I said to Dick's gentle, elderly father, "very kind of you to let us use your dock. I'd feel better about it, though, if you would let me pay you."

"Thank you, but no. It isn't necessary, not at all. So long as there's room for you, you're welcome to it. On these islands, you know, neighbors have to depend on one another." I wondered precisely what he could depend on us for, this brand-new family of greenhorns that had just moved in next door. "And there have been children playing around that dock for fifty years," he continued. "Seems natural to have yours there. The only thing is, I'd like them to be a little careful not to leave things lying around cluttering it up."

"We'll see to *that,*" I lied—knowing how we would warn, scold, urge, and fail. I tried to remember where I had left my own fishing rod and couldn't. "You fish, Mr. Johnson? Maybe we could go out together sometime?"

"Never owned a rod," he replied, "and neither has Dick."

"And you've been coming up here for fifty years?"

"Longer than that," he said.

"Then . . . then . . ." I stammered, "excuse my asking, but if you don't fish? . . ."

"You mean, if I don't fish, why do I come?" He laughed. "Well, I guess I just think it's beautiful here, that's all." I thought of our breakfasts on the back porch. The house protected us from the prevailing southwesterly winds, and the morning sun plated everything out back with warm gold. However early those breakfasts were, we could always hear the sound of hammering or the rhythmic hiss of sandpaper from the Johnsons' shed. He was always fixing a chair or refinishing an oar.

"Always something to do," he added. He understood the pleasures I was only beginning to discover—the happiness of coping with problems oneself, not hiring someone else to do the job. Maybe it was the island that had taught him. It was certainly the island that was beginning to teach me.

"Judging from the noises I hear from your shed, this isn't much of a vacation for you up here."

"I took a vacation once," he said with amusement, "and I'd never do it again. They're probably all right for some people—the ones that like just going somewhere and looking. By the way, I notice your bowline is unraveling."

"Ah, yes. I meant to put a knot in that," I said guiltily.

"A knot would do," he admitted, "but if you'd like to step out to the shed where we've got a workbench, I could show you how to splice a line back on itself. Makes a better job, and looks nice too." I went with him and learned. Next day when we went for groceries, I tied up at the town dock with new pride in lines that were neatly spliced back.

While in town, we stopped at Mitchell & Wilson, the lumber and construction company. I asked if they could send someone out to give us an estimate on what it would cost to build a dock. They said they would do that right away, and a week later they did.

We had a stretch of ancient catwalk that began just before the willow tree and dithered on its shaky supports forty feet over rock pools to end in mid-air three feet above a pile of rocks that ran another twenty-five feet to another huge outcropping of bedrock perhaps thirty feet square. We called this outcropping our shoal, but it was more like a separate little island. Some storm of bygone years had tossed a great tree trunk up across the front of the shoal, making a fine perch for fishing. The river face of the shoal dropped almost straight down into sixteen feet of water. It was a wonderful spot. My only problem was to keep from breaking my neck while getting to and from it.

"That shoal . . . I think that shoal would be an ideal place for a dock," I told the chap from Mitchell & Wilson. Much practiced, he jauntily picked his way over the boulders while I, with some embarrassment, took almost to all fours—a demeaning posture, but it kept down the hospital bills. "Plenty of water there," I said, scrambling after him. "Used to be a dock out here, my neighbor said, a great big hefty platform."

"I can see that," he observed.

"Perfect spot, eh?" I said. He was silent. "Don't you think so?"

"Like you said," he sighed, "there used to be a dock here, Mr. Russell. Still some of the rods coming out of the patches of concrete." He paused, then appreciatively patted the wave-polished tree trunk. "Nice timber here."

"Good spot to fish from," I answered, sitting down on the trunk. "Good deep water. That's why it would be such a great place to have a dock. Don't you think so?" He said nothing.

"Tell you," he said finally, sitting down beside me and lighting his pipe, "be no trouble putting a dock here. We could probably start next week."

"Fine. . . . How much do you think it would cost?"

"I don't know," he said casually, "maybe a thousand . . . maybe fifteen hundred." I took the blow in silence. "But the thing is," he continued, "the thing is . . . well, you pointed it out yourself . . . there used to be a dock here, a big one, but it isn't here now. See, you get a lot of weather here," and he patted the trunk. "Must have been a good blow that tossed this one up here. You get the full blast of the southwesterly winds here, and that's where most of our storms come from. You've got those waves coming in from maybe ten miles of open water right in from the Forty Acres. Now, to put anything here you could tie a boat to, you'd have to have a pretty good overhang, because you don't want the boat pounding up against the rock. There are some pretty sizable points sticking out too, just under the surface," he continued. "Well, with that overhang," he went on patiently, "it means those waves'll be coming up right under your dock, and they'll hammer it up to beat hell. I mean, suppose this tree here had drifted under your dock, and the waves started slamming that up under your overhang!" Like a battering ram, I thought.

"But . . . but isn't there any way? . . . I mean, there must be some way to make the stringers firm," I said weakly, not seeing myself how anything could be made firm enough to withstand uppercuts from a tree trunk.

"That's what that cement's all about," he explained kindly. "They cemented those half-inch rods down. They came up through heavy stringers and were bent over like hooks at the top so as to hold those timbers in place. Now, just you come with me and have a look at this one over here." I went with him to where a rusty shaft rose from a concrete foundation. The rod was vertical. "Look there," he said running my hand down the rod. "Like I said, that used to be bent over at the top. It made a sort of clamp, and a good clamp too. That went right through a timber. Can you guess how that rod got straightened out? I'll tell you. The timber was hoisted right up that rod and straightened it out just as if it was a hunk of baling wire."

"Guess so," I admitted.

"It isn't just waves," he went on. "There's ice. She ices up pretty good in the winter . . . maybe twenty, sometimes thirty inches thick. That's all right; it's just when it breaks up in the spring. You get a piece . . . maybe five or ten acres of it . . . driven along by a pretty good wind. That piece of ice doesn't move all that fast, but it's got momentum. When it hits something, that something's gonna move. I've seen whole boathouses just kind of explode under the pressure. Like somebody put a bomb or something under them and blew them into kindling." As he talked, I pressed my foot against the shoal to make sure it was still there.

"If that's the case," I said, "how does any dock last up here?"

"Good question," he chuckled. "Luck more than anything. Most people build their docks on the lee side of the islands when they can. Protected like that, they can last a good while, but then you get a storm, the odd one out of the east, at breakup, and that's the end of that!"

"But, damn it!" I exploded, "there's my neighbor's dock only seventy-five feet away, and that's been there for fifty years!"

"Don't doubt it," he agreed. "Looks like we might have built it."

"Well, then, all I'm asking is that you build me one like that! Fifty years is good enough."

"Can't be done," he returned. "The thing is, you see, your shoal here is his protection. This sticks out thirty or forty feet beyond the end of his dock, and he's downstream. Ice comes whacking in here and bounces off right on by him. He's snug as a bug, but you're not. Oh, I can build you a dock if you say so, for maybe fifteen hundred, and she'll be a dandy, but I can't promise you it would be here next summer."

"You mean, the long and short of it is that we just can't have one?"

"You can, like I said. You can have a new one every year, but we don't need the business that bad." I followed him back to the shore in despair. As he untied his boat, he asked, "How do you get on with your neighbors?"

"How do you mean?"

"Do they mind your tying up here?"

"No. They're very nice about it."

"There's your answer, then. Keep on the good side of them. Course, we'll do whatever you say, but I know what I'd do." He yanked the cord and his motor buzzed.

"Yeah. Thanks for the advice."

"Better still, take it. So long," and he putted off out into the channel.

It was good advice, but it wasn't what I had wanted to hear.

"Nice of him to be so honest about it," Elisabeth said sadly, "but I do wish . . . maybe sometime . . . but right now, how about those thorn trees? They're an absolute menace!"

Throwing on a jacket for protection, I grabbed the hatchet and went angrily to work.

"What's so important about a dock anyway?" asked a mainlander whom I was boring with my problem.

"Would you buy a house with no place to park your car nearby?"

"Maybe," he defended. "I could always walk."

"It's kind of hard to do that on water," I pointed out. "In town you can choose between walking and driving. On an island you've got no choice. You either take a boat or stay home."

But a dock was more than a convenience or even a necessity. The Johnsons gave us those. But the need for a dock of our own was deeper than physical necessity—it was the need not to have to say thank you. I needed to be able to say to Dick and his father, "Tie up at ours any time. There's plenty of room!" But we didn't have an "ours." Only a shoal. Not being able to say that to them bothered me, and it bothered Elisabeth. And Richie and Mark swam off their dock, and they were always leaving sneakers, shirts, or towels down there, no matter how often we lectured them. Then, too, with children, swimming and yelling go together like crackers and cheese, and we were always afraid they were disturbing the Johnsons. More than all these was the need to feel that, when we touched shore, it was our shore. Home port isn't home port when it belongs to somebody else.

"Cement. Cement's your answer," declared Durrell Enck over a beer one winter night. "If you've got a rock base—wish I could see it!—why, cement would be there till kingdom come!"

"Bet you're right," I said, as the truth of what he said came up, like slow dawn. "By God! I bet you've got the answer!" Then I thought a bit and the dawn faded. "But," I mourned, "you can't pour cement on empty air, and this dock's got to have an overhang—maybe three feet—because there are big chunks of rock that stick out under the water that would grind holes in any boat moored next to them."

"Hell with overhang!" he cried with abandon, banging his empty glass on the bar. "Let's have a couple more here, bartender."

"That's easy enough to say, but saying doesn't get rid of those points. They're like big granite teeth! There's got to be an overhang!"

"Hell with the teeth," he said confidently. "We knock them off, that's all." With his confidence and his use of the "we," my heart began to pound. "Blast," he continued, "that's all. Dynamite's a great invention." Nervously I fingered the fresh cold beer, trying to imagine what a plunger must look like. Everything I had read about blasting always seemed to have what they called plungers that you pressed in. "Three, two, *one!*" What was the powder monkey's equivalent of timber? With a trembling hand, I raised the glass. "No sweat," he said carelessly.

"Durrell says," I told Elisabeth, "that cement's the answer . . . cement and dynamite." She thought about it for a minute, then said, "I hate to say this. . . ." I prepared myself. Whenever she started that way, she invariably went on with special relish, and what followed, while always unpleasant, was sometimes even true.

Then came the *"but . . ."* "I agree we need a dock, all right, but worse than a dock, what we need is a decent boat and motor. I mean, that boat of ours would be just great for some moldy little pond or other, and as for that, that con*traption* on the back end —that so-called motor—that, that great *buy* of yours! I mean, it's going to be precious little consolation to go out onto the river thinking that our bodies could be washed in on our own *dock!* Remember Auntie Em's lines: *'Save the children!'* Well, every time I get into that boat I feel like saying, 'How about me too? I could do with a little saving too.' The fact is, we all could." Beginning as she did, she threw me off balance because nobody could like shopping for boats and motors more than I.

"Sure," I said, recovering. "Right! That's exactly what we'll do. Next summer we'll hunt around, and I bet you we can pick up . . ."

"Oh, no you don't!" she said, reading my thoughts. "None of your great buys this time! No more of this secondhand business! They're both going to be brand *new!* There're going to be no leaks, and the motor's going to *work!* Actually *run!*" She was under a real head of steam, the kind that charges right through price tags, so there was no point my saying anything. Besides, the notion of stepping down into a dry boat, pulling a cord and having the motor start, and having it actually go on running until somebody shut it off—all that sounded pretty good to me too. I just didn't believe you could buy those things, that's all, but I didn't say anything.

With her degree and year of graduate work from Oxford, Elisabeth was eminently qualified for teaching. She taught a section of composition in our night-school division, and luckily I had one too. Both meant extra money. Where she found either the time or the energy I can't explain, but she taught with the same passionate enthusiasm and zeal with which she set about cutting the grass or doing anything else. Her checks chipped away at the bank loan, and with my extra, we established our Island Life Improvement Fund. (Those were the days, 1956, before the acronym came into its own.) The first target for ILIF was to provide us with safe and reliable transportation.

I walked home through the dank chill of winter twilights, my heart bright with the possibility that the mailman had brought another catalogue that day. Next to and in between discussions of Chaucer, toilet training, and the heating bills, we pondered on the length of some model or other, its beam, and the depth amidships. After months of painful dithering, we settled on a sixteen-foot boat constructed, or so the catalogue declared, out of a new, revolutionary material "stronger than steel"—something called fiberglass. Though deeply uneasy about the "glass" bit, we nevertheless screwed up enough courage to send off the check to Herters in Minnesota with the order. After getting clearance from my sister Hunnie, who lived in Syracuse, New York, we asked that it be delivered to her address. The rest of ILIF went to a marina in New Jersey for a new last-year's model of a fifteen-horse Johnson.

The motor came to Lancaster. I carried it carefully upstairs and put it down beside my desk, where I could be able to reach down and pat it every now and again.

"Looks great, just great," said Elisabeth excitedly, "but how is it supposed to work? I mean, it doesn't look like our old Elgin. What are all those levers and things for?" Soft-pedaling my complete ignorance, I came on strong with confidence about its all being very easy once one got the hang of things. I closed by calling Richie and Mark to come and admire, which they did.

"But where's the boat?" asked Richie.

"Your boat arrived today, Rob," said Hunnie's voice on the phone early in May, "and is it ever big! It looks like a barge sitting out there in the back yard—big enough to take the whole family, what's more." That meant something coming from her, with her five daughters. "Tom can hardly wait to bring it up to the island. . . . Only thing," she said a little nervously, "is that . . . well, you can see right through it. You think that's all right?"

"Supposed to be stronger then steel," I quavered.

A month later I sat on the Johnsons' dock massaging the motor and worrying about those levers. The great day had arrived. Tom was scheduled to appear shortly at the Bay Street Dock with our boat in tow, and John Deacon, the son of a good neighbor, had offered to run me in with the motor, wait there, help launch this strange glass boat, and guide us back in his safe, dependable wooden craft. He putted up to where I sat and we loaded the motor.

"But is it really going to work? That's the question," gloomed Elisabeth from above us.

"Just hang on to that guarantee," I called back doubtfully as we hummed off.

"Actually fiberglass, you say?" John inquired as we waited in the parking lot.

"Yep," I answered. "Fiberglass."

"Well, I heard about this fellow in Toronto. He was unloading a fiberglass boat from a truck. It fell or something. Anyway, it broke and cut his arms to shreds."

"Not supposed to," I said.

"That's what I heard, that's all." Nervously, I lit my pipe.

"We'll be real careful unloading it," I said.

"Guess we'd better," he agreed. "You say you can see right through her, eh? Must be really odd. Sure am glad I'm actually going to see one, but I guess I'm glad it's yours."

"They said it was big, though," I added, "very big." Gravel hissed and popped under the tires of an approaching car.

"It's him," John cried, getting up. "Wow! It does look big, all right."

"There she is," said Tom triumphantly after the introductions. "Look her over." I did. It seemed vast. Gently, I tapped the gunwale with a knuckle.

"Go on," called Tom, "hit it! You can't hurt it. This fiberglass is harder than metal."

"You selling them?" John inquired.

"Wish I were," Tom replied. "Come on. Let's launch her."

"Carefully," I advised.

Tom jumped back into the car and backed the trailer down the launching ramp.

"Well, I'll be damned," John cried, "it floats—fiberglass floats! But hey, let's put a line on it before it floats away."

Under his seamanlike direction, the "careful" launching was accomplished, the motor put in its place, and everything readied for her maiden voyage. During the whole operation, with the three of us climbing around inside her, she kept as stable as a truck bed. As for saving the children—they could sit on the gunwales of this craft. She could take all of us with plenty of room to spare. The dog could be as restless as he liked. The catalogue had declared her to be a sixteen-foot hull with four aluminum seats—a utility model, and that's exactly what she was. Nothing fancy—no deck, no windshield, no plastic-covered foam-rubber comfort—just plain, hard, roomy, solid fact. With its five-foot beam and thirty-inch depth amidships, and with those four broad seats, there was room not only for all of us, but for Tom and Hunnie and their whole flock as well.

After a couple of yanks on the starter cord, the motor spoke, and not in a series of hoarse, resentful, intermittent barkings like the old Elgin's. Its voice was a soft, deep-chested growling, confident, resolute, powerful. It sounded so good, I couldn't believe it could be ours. Its voice declared it to be the sort of equipment that belonged to someone else—to any one of those other people tied up at the town dock who, their errands done, stepped carelessly into their boats, secure in the knowledge that the motor would not only start, but purr happily all the way home. But no! Not theirs; *ours!* Now we could publicly exude that sort of casual assurance we had so envied. Now the tables would be turned: now we could zoom over to some poor devil who sweated, swore, and raged over a silent engine as he drifted helplessly down the channel; we could sidle up to him with: "Can we give you a tow anywhere?" That's what being rich must mean, and, by God! we were it!

John returned to his boat, and Tom followed him back to Hay Island through the darkness. They had heard us coming, and everyone was waiting on the end of Johnsons' dock armed with flashlights to inspect this mad investment of ours—this fiberglass. Over the murmurs of wonder came Elisabeth's laughter; "Good Lord! good *Lord!* it's enormous . . . just e*norm*ous. There's our year of night-school teaching, and it floats!"

"Sure does," admitted John from his boat, "but if I were you, I'd be kind of careful with it. Glass, you know! Good luck," and he started for home.

"Agh!" Tom threw after him, "you could take this out in a *hurricane!*"

Four nights later: "Bob! *Bob!*" Elisabeth's voice punctured the balloon of my train trip. However, as I roused to the familiar lumps in the mattress, the vibration of the train remained, as did its not so muffled roaring. "Bob! The *wind!*" Wind? Could that be wind? Unprotected as we were except for our willow tree, the wind often moaned, whined, and sometimes even howled through the cracks around the front door and about the corners of the house, but this was no eager, inquiring husky out of Sergent Preston of the Yukon anxious to come in by the fire. This was too big for any animal. The whole cosmos was doing calisthenics. This was the kind of power I had sensed nineteen years earlier behind the canvas windbreak as I lay on the hay, listening. "The windows!" she urged. "Sounds like it's going to blow them right out of their frames!"

"Hell with the windows," I said, hunting for my shoes. "I'm wondering about the house!"

"But the boat! What about the boat? Hadn't we better go see?"

"If there's anything left," I groaned, "I suppose we ought to give it decent burial."

"A year's work," she almost sobbed as I pulled on a jacket. She was already dressed. As we started downstairs, something heavy crashed and thundered across the front porch.

"Probably a chair," I said, reaching for the handle of the front door.

"Hope it wasn't the remains of the boat," she said. "But don't open that! We'll never get it shut again!" We tried the side door. Because of the pressure built up by the wind through the cracks in the front of the house, the door was hard to open. Once open, it flung us out and shut itself, slamming behind us. The side porch was like the back edge of an airplane wing, a sort of partial vacuum created by the corner of the house. As we went down the side steps and out of its protection, the full blast sent us staggering.

"Whoa!" Elisabeth cried. "Hang on!" Clinging together and bent almost double, we plodded through what seemed a solid wall of air the hundred feet to the dock.

"Can you see the boat?" I shouted.

"Think so," she called back from beside me. "Yes. Still there."

"Should have brought rope!" I shouted.

She yelled back that she would get some and disappeared. I dropped to my hands and knees and worked my way along the right side of the Johnson dock the fifty feet out to where our boat reared and plunged. With each wave the mooring lines sang like violin strings, then relaxed, only to snap taut again. I found the gunwale with my feet and dropped in after the next wave. The boat was half full of water. Had it poured in through some gaping hole or was it only from the wave crests over the bow? There was no way then of making sure, but I started for the stern to fish for the jam tin we had jokingly dropped in for a bailer. The boat rose and backed on the next wave and came up short on the moorings. The jolt sent me sprawling over the stern seat, and for a moment I lay crumpled and sloshing around with the gas tank and the jam tin.

"What are you doing that for?" Elisabeth shouted from above. No time to point out that I wasn't doing *it; it* was doing me.

"Trying to bail this thing out to take some of the weight off the lines," I shouted, "but this tin's no good. I need a *pail.*" There was no answer through the roar of wind and water. Steadying myself, I furiously plied the tin.

My weight added now to the weight of the water, I reflected. Could be enough to break the lines, and I would certainly be washed away. I wondered where and when they would find me. "Never mind," I commanded myself sternly, *"bail!"*

"Pail!" she shouted from overhead. I reached up and took it as another wave hoisted the bow, driving the stern back and down till the stern line pulled it up. As the stern rose, the wave crest poured in over the bow.

In the heaving, thrashing boat, I felt like a rodeo rider on his bucking bronco. They certainly deserved respect for being able to keep their seats, but I wondered how well they would do if they had to bail at the same time. With the bucket, though, I was able to make some progress. As I did, the fiberglass lightened, rode higher, and took less water from each wave. Maybe she wasn't stove in, then. But as I gradually freed her of the burden of water, her riding of the waves became more frenzied and time and again I was thrown from the seat into the bottom.

"Come on," Elisabeth finally cried. "That's all you can do. *Here!* I've got four ropes tied to rings in the dock. Tie the other ends to the boat." Cautiously I made each line fast—cautiously, because a finger or a hand could easily be smashed if caught in a loose knot suddenly tightened by a dive of the boat. The last line secured, I leaped onto the dock between waves.

"Don't see what more we can do," I shouted. "How does it look?"

"Looks *wild!*" was the answer.

"How's the Johnsons' boat?" I called.

"Doesn't look bad," she replied. "It's on the other side of the dock, and that side's protected by our shoal." For an instant I thought of trying to move our boat over to that side, but only for an instant. To try to move it would be to lose it. "Come on," she continued, "or we'll get blown away!"

Driven by the wind, we returned to the house at a staggering run. The back door slammed behind us, and we were safe in the kitchen. Or were we? Was anyone?

"Two-thirty," she said. "Might as well go back to bed. No point in sitting here. I'll get you some dry pajamas." It couldn't get any worse, I thought, so we went to bed and even fell asleep to the roar of the wind and the rattle of the windows.

But it did get worse. When I woke, the house was trembling under the steady thunder of the wind. But, I reflected, as long as it doesn't actually collapse, we might as well get on. I built a fire as Elisabeth fixed breakfast, and we all ate it on the floor, huddled around the cheerful blaze.

"Is this a flood?" Richie inquired fearfully.

"No," I reassured him. "Heck, no. You don't get floods on the St. Lawrence."

"Why not?" Elisabeth asked skeptically.

"It's too big," I explained, "much too big."

"How about the Mississippi?" she pursued. "I thought that was big, and that floods, doesn't it?"

"That's different," I declared.

"How?" was her simple question.

"Well . . . well, you see, there are lots of rivers running into the Mississippi. They bring lots more water . . . all the snow that melts along all those rivers. It's what you call a watershed, a huge watershed!"

"Mm," she mused doubtfully, "but didn't you say all the Great Lakes emptied into the ocean through the St. Lawrence? Doesn't that make this a big watershed?"

"Damn it! This isn't a flood. The St. Lawrence is just too big for a flood. It's thirteen miles wide here. Where is the Mississippi that wide? Tell me *that*."

"I thought down by New Orleans or somewhere it was things like fifty miles wide, and *they* . . ."

"Sure. Right. They do, but this is different. Whoever heard of a flood on the St. Lawrence?"

"Well . . ." she replied, "do we have to have heard of one before we can have one?"

"Never mind," I answered. "How about going on with our book?" She agreed, and Richie and Mark scrambled up to find it. So the morning passed with Peter Rabbit again undergoing his perils in Mr. MacGregor's toolshed, Tommy Brock, the badger, invading the fox's house, and I tending the fire.

Hay Island, that hundred acres or so where there would be "lots of room for the kids to play and a good supply of neighbors," had shrunk to an eight-foot semicircle before the stone fireplace. For me, to step across that magic line was to venture into chaos. As I did so to fetch wood, Elisabeth's voice was instantly swallowed in the vast tumult of the wind and the shudder of the house as it seemed to reel under the weight of a sudden and even more powerful gust. Though the woodpile was protected by the house from the full blast, the pieces of driftwood I pulled out seemed alive. Each piece thrashed in my hands as if some angry demon had hold of its other end and was determined to pull it from me. If something happened, I reflected, Dick and his father might as well be back in Philadelphia, for all the help they could give. Asking Dick to lend a hand with the pump was one thing, but it was quite another to ask him to help hold the roof on or the cottage up, especially when his roof and his cottage might come apart first. And I could think of nothing to do in the way of prevention. All I could think of was to wait and see what happened, and, when things quieted down, to clean up and, if possible, repair.

"Tell you what," said Elisabeth, "how about some brownies? I'll make some brownies. That ought to cheer us up." She rose, and I shouted after her as she disappeared into the roar: "Have another look at the boat!" She could just see it from our dining-room window as it rose on a wave above the dock next door, and periodically she had checked to make sure it was still there. If she answered, I didn't hear, but I felt sure she would look; so I busied myself with the fire and giving Mark some sawing lessons. We had three or four good lengths cut before I began to wonder where Richie was. He could have been singing and skipping rope on the other side of the room and I wouldn't have heard him.

"Where's Rich?" I asked Mark.

"Richie there," he declared with two-year-old precision.

"Show me," I said, so he took my hand and led me into the dining room, where Richie stood by the window. Taking his shoulder, I drew him back, explaining that it could blow in on him. He came back with us to our stack of wood by the fireplace.

"You been helping Mom with the brownies?" I asked him. He said he hadn't. "Okay. Then listen, you two. Stay here, but don't put anything in the fire till I get back. I'm just going to see if I can give her a hand." I waded through the tumult to the kitchen and shouted for her. There was no answer. Again; still no answer. A little concerned, I searched the house, going into each room and shouting for her. No success. My concern blossomed into worry as it occurred to me that she might have fallen and could be lying unconscious somewhere. Unless I happened to kick her, I might never find her, so back to the fireplace for eyes.

"Richie," I cried, "would you come help me find Mom?" Calmly, he took my hand and stood still. "No. We've got to look, got to *find* her!" Confidently, he led me into the dining room and stopped before the big window. "No, no," I tried to explain. "She went to the kitchen. Make brownies. Remember?" and I tried to lead him toward the kitchen. "But I can't find her and you must try to help."

"Boat," he replied simply.

"Boat? What boat?"

"Down at the boat," he answered. Then it hit me. On her way to the kitchen she must have seen something and dashed. I had never even heard the door slam behind her. And that was why Richie had been standing at the window. He had been watching her.

"Why didn't you tell me?" I shouted as I dragged open the side door. "Take care of Markie and don't touch that *fire!*" The wind slammed the door and drove me down the side steps. I was out in the tumult, the hissing, rushing roar. To my left was the river from which came no periodic thunder of waves—only one prolonged, continuous crash. Bent double, I fought my way toward that wall of sound, my feet searching for familiar ruts and rocks, my ears straining for the sound of a voice that was not there. I shouted. But had she been ten feet away, she could scarcely have heard.

I struggled toward the river, angling to my right where the two cedars ought to be and between which we passed before turning left onto the Johnsons' dock. Would they still be there? Would the dock be there? Pellets of spray stung like buckshot, and then the branch of a cedar lashed across my face. At least one was still here. Making my way around the anguished tree, my feet found the narrow rut that was the path to the dock. Down, down, down the path till now I was inside that solid crash of wind and water; then the smoothness of the first heavy plank. Dropping to all fours, I found the right-hand edge beside which the boat had been moored and crawled out farther into the turmoil. Beneath the heavy topping, the four-by-eight timbers trembled. The whole structure shuddered as each wave exploded over the seaward end. I kept crawling, kept shouting, kept listening. Then, the first mooring ring. No rope. Had we tied this far in, or was it farther out? I kept crawling.

"Down here!" She was in the water. Gripping the edge with both hands, I swung over the side, hung for an instant, then dropped. As I hit the water, a wave sent me sprawling. I scrambled to my feet. It was only a couple of feet deep, but the rocks were slippery.

"Out *here!*" she called. The water was strangely warm and soft about my legs as if it were a denser form of air. Another wave sent me staggering back.

"Here!" she repeated. So as not to take the full force of the waves, I sidled step by step toward her voice.

"This way! . . . *this* way! . . . Out here! I *can't* let go!" Then I was at the boat. The gunwale was low in the water. The boat must be almost full. Grabbing it as hard as I could, I shouted: "Hell with the boat! Get up on the dock!"

"Ropes broke!" she shouted back. "Hang on. I'll get help!" Help, I thought? Where? But her voice had disappeared. Thigh-deep, I stood at the stern trying to point the bow of the sixteen-foot craft into the next wave. Had she been empty and riding high and free, this would have been hard. Heavy and sluggish as she was, there was no hope of controlling her. I felt the bow rising and braced myself against the rocky bottom. Its backward surge knocked me off my feet. Still clinging to the stern, I went under, the boat plunging back over me, and a rock bumped along my back as I was dragged shoreward. I pulled myself from beneath the boat and stood beside the motor, whose lower unit rose and fell with the motion of the boat like a guillotine or meat cleaver. To get a foot or a leg caught between that shaft and the rocks below would not be a good idea. If I could just tilt the motor up, I might get rid of the threat of the sledge hammer rising and falling along my right leg. To do this, I would have to find the release mechanism—one of those strange levers Elisabeth had asked about—and to hunt for it, I would have to let go at least with one hand. I risked it, found the lever, pressed it down, and fought the shaft up to the horizontal. Then another wave drove the wallowing hulk back over me. Still hanging on, I again pulled myself out from under it. How in God's name had Elisabeth managed, I wondered—and for what must have seemed hours!—while I had quietly been helping Mark with his sawing.

How stupid could I get? She hadn't been at the stern! Nobody could hold that thing back there! Now it was sideways to the seas and shipping more water with every breaker. Fighting along the gunwale to the bow, I clung to that and let the waves straighten her out. If I could hang on with one hand and bail with the other . . . but where was the bailer? To find it, I would have to climb in and fish around, and by the time I found it, where would we be? No. I would just have to do my best to keep her off the rocks.

Left side to the oncoming seas, shoulder lowered to meet them, feet braced, I clung to the bow that heaved and swung like the trunk of a moody elephant. Each wave sent me reeling shoreward, and, before the next, I struggled to get back the ground I had lost.

Suddenly it struck me: this was all absurd. It couldn't actually be happening—certainly not to me. It couldn't possibly be Robert Russell spluttering, gulping, gasping, slipping, sliding, being dragged, heaved, and knocked about by this peevish monster—not really *me* trying to control it with my bare hands! I was a professor of English—my function was to stand before rows of silent undergraduates explaining why Robert Browning was a "major English poet. His dates were . . ." Splash! gurgle! bang! Up again, not even being able to rub the bruises! Never mind the bruises. Back out again before the next one catches me. How did I get here, anyway? What would those undergraduates think, what would they say if they saw that "professor" here, of all places?—that same professor whose shoes squeaked in the tension of the silent classroom—was it? yes, surely that was the silence of tension—as he shifted his weight, struggling to explain what he actually thought about that "Last Duchess." The right arm yanked almost out of the socket, feet slithering, knees buckling, being bounced over the rocks; up again and back out. No shoes here, no classroom; no students; no Browning. Only the here and the now with the heave of waters in the roar of a world gone mad!

How long was that struggle? Long? The wrong question, because it asks only half. Time is vertical as well as linear. Time is the medium in which experience occurs, and the height of an experience is usually more important than its length. Its height is the extent to which our consciousness rises to sense the full import of the moment. Beside that dock, thigh deep between waves, submerged by their crests and clinging to the boat, the height of the struggle was clear because it was so simple: "Can I survive?" Then why not just let go, let the boat be smashed on the rocks, find shore, and make my way to the house where the children were alone with a blazing fire? However sensible, it had not then seemed even a possibility. Though I had told Elisabeth to let go, to get to safety, that had been before I had taken hold, before I had accepted the responsibility. After having taken hold, letting go was impossible. The boat and motor were perhaps already badly damaged, maybe past repair. I was hardly in a position to give it careful examination. But, whatever their condition, they were something we had worked for and won. They were not to be abandoned. It was that which had kept Elisabeth firm at the same task. For the same reason, I could not let go. This thing that yanked me up and off my feet, that dragged me bumping over the rocks, was ours and not to be torn from us. It wasn't dollars and cents to which I clung, but a significant advance in our struggle, our planning and efforts to make a foothold, and then a way of living on this island and on this river. To let go would be to say that it had not been worth it, and it had been and would be.

"Bob!" she called from far behind me—I must have been washed along the shore—"Tom and Tommy Deacon are coming."

"Great," and I began dragging the boat toward her voice. The Deacons lived on the other side of the island. Tommy was fifteen, strong and river-wise. The four of us ought to be able to manage.

When they arrived, I held the boat as they removed the motor and carried it ashore. Then, after bailing as much as we could, we heaved the hull up onto the Johnsons' dock, then dragged it to safety under the cedars. Exhausted, we sank on the grass beside it.

"We've lost ours," said Tom sadly. "The side was stove in with the waves smashing it against our dock." I commiserated. "And this is that fiberglass, eh?" he continued, rapping the hull. "Can't see so much as a scratch on her anywhere."

Tommy had been examining it carefully and announced, "Looks as good as new to me. Must be really tough stuff."

"Tell John, will you, that it doesn't shatter." Tommy laughed.

"I don't know about you," put in Elisabeth, "but I could use a good cup of tea." We trooped back to the house where the children had not been playing with the fire.

"We've got a couple of shipwrecked mariners over at our place," Tom observed over his first cup. "Washed in late last night. Their boat was smashed too. Lucky pair of chaps, they are, to be alive."

I made a stab at thanking the Deacons as they left. "No, no, my dear fellow," Tom declared in his English accent, "not a word. We've got to stick together on these islands, especially in hurricanes."

"Hurricanes?" asked Elisabeth.

"Yes, hurricanes," he repeated. "Heard it on the wireless before the power went—hurricane Eileen, Hazel, or Louise, or some such."

"Imagine that," I said. "So this is a hurricane. Richie thought it was a flood."

"Oh no," Tom laughed, "we don't get floods, not on the St. Lawrence."

When they had gone, Elisabeth groaned, "Brand new—not even a week old! and ruined by a bloody hurricane."

"Don't worry," I consoled, "they're all right. Anyway, when it calms down we'll go to town and take out insurance on the whole outfit."

"That's just typical," she exploded. "Just *typical.* We won't need insurance when it calms down."

"Cheer up," I urged. "Maybe next week we'll have the flood."

5

The Dock

"Cement, eh? So you think that's the answer, do you?" said Jack, banging his empty glass on the rickety table. "Hey, Skip! Three more draughts here. Russell's got some American money he hasn't spent yet. If cement's your answer, then I'm your man. Isn't that so, Indian?"

"That's right," came the soft voice of Howard, a full-blooded Iroquois. Skippie plunked three beers on the table and took Jack's money.

"Here," I called, "this one's mine."

"Too late as always, Russell," Jack laughed.

"It's my turn," I objected, "and anyway, this is my business conference, isn't it?"

"We'll get it out of him, won't we, Howard?" Jack said, pouring salt from the large shaker into his beer.

"We will if we can," Howard agreed. "Jack here's your cement man, but you need a powder monkey too, you say. Well, that's where I come in. I'm a powder monkey."

We were sitting in the men's bar of the International Hotel, Gananoque's hangout for fishing guides, painters, handymen, and general river rats. It was the lowest joint in town by a long shot and, by an even longer shot, it was my favorite. The two or three other places were cleaner, quieter, and more respectable. They catered to the upper class and tourists, but those places didn't smell right or feel comfortable. They lacked contingents of regulars who could always be found on tap, any of whom would talk river-talk over good, rich Canadian ale till it was long past time to be home. The atmosphere in the International was a special combination of beer-sodden wood, camaraderie, and the peculiar cheerfulness born of common hardship that is determined to enjoy life anyway. Skippie ran the place with good humor and more than a little tolerance. When things got a bit rough, as they sometimes did, and a chair or two was smashed or maybe a little paneling knocked out, it never bothered him. He knew the culprits would be back the next day after work with materials and tools to fix things up good as new. I asked Jack how Skippie could be so sure things would be set right.

"Well, it's our place, isn't it?" he answered. "Look at it this way: If we don't take care of it, where the hell else are we going to go for a beer? You wouldn't catch me or the Indian here up at the Golden Apple, for Chrissake! Whatever Skip wants done, he knows he's only got to say the word and there's guys that'll do it. He treats us right, so it's only fair, eh?" And it worked. "But about this here cement job. It's going to cost you a pile, you know." I recalled the "thousand or fifteen-hundred" that had rolled so easily off the tongue of the Mitchell & Wilson man.

"Been thinking about that all winter," I said, "and shoving the odd dollar in a sock. But first, what I need to know from you guys is whether the job can be done."

"Hell yes, it can be done. Do anything if you've got enough of them Yankee dollars stashed away. Christ, I been pouring cement since I could see under a dog's front legs, and Howard here's the dynamite man. You heard him say so himself. Leastways I heard him say that. Don't know whether I believe him, but he said so. No shit, Howard, you really know anything about powder?"

"Christ, yes! Course I do," the Indian complained, his dignity wounded. "Why, didn't I knock out a great boulder down here at Bloodletter for that Captain Blowhard or whatever his name is?"

"I don't know," Jack said. "Did you?"

"Course I did," Howard declared. "Thirty-seven sticks, it took. We had one helluva time!"

"But this is under water, Howard. Will it work under water?" I asked.

"I don't see why not," he said slowly. Then, "Why, Christ!

What am I thinking of? Sure it'll work under water. They blow plenty of things under water."

"They do," laughed Jack, "but the question is: have *you,* Howard?"

"Jesus, Jack! Everybody's got to learn, don't they?"

"You guys seem pretty confident," I said, "but you haven't even seen the place where I want to build this dock. How about jumping in the boat and having a look?"

"What the hell do you mean, I don't know it!" Jack exploded. "I been screwin' around this river for thirty-five years. It's that lousy little pile of rock out in front of your place, isn't it?"

"Maybe you know where it is," I argued, "but you haven't really looked it over, not carefully. Come on!"

"Sit down," he said firmly. "What's your hurry? We aren't finished drinking. That shoal's been there a long time, and it's not going to float away while we have a couple o' draughts. Hey, Skip! Three more."

"If you know it so damn well, then how much is it going to cost me to do the job?" I said, sitting down.

"Depends how much you got," he said.

"Well . . . I've screwed together . . . saved, that is . . . oh, around three hundred, maybe."

"Holy *Christ!"* Jack roared, leaning back, "it's not going to cost *that* much."

"Well," I hastened, "I didn't mean *three* . . . not *three.* I said two, didn't I? I said two hundred."

"That's better," said Jack relaxing. "That's more like it, isn't it, Indian?"

"We'll *take* three," Howard admitted softly with a tentative laugh.

"Enough's enough," Jack said. "How about it? I know where I can get a mixer for the concrete, but we're going to need a scow to get the cement and gravel over there. Got any ideas?"

"My brother-in-law's got one he borrowed. Got a leak or two and it's none too big, but I can get that."

"Then that's settled. Good. Take us a day or two to get things lined up. Just relax, Russell. Howard, here, 's got to dig up the dynamite too."

"Don't you need a license for that stuff?" I asked.

"Let's put it this way," Howard said. "I'll get the powder."

"But how about the license?" I pressed.

"Augh, hell," was Howard's only comment.

"Tell you what, Russell. You just lay in a couple cases of beer," Jack advised, "and we'll see to the rest. License! Christ, ain't nobody going to give the Indian no license, but that's our worry, not yours." Then, softly to Howard, "You really know somewhere you can get it?"

"You can find anything if you look hard enough," was his simple reply.

"You really think they know what they're doing?" Elisabeth inquired nervously a week later. "I mean, the dynamite and everything?"

"I hope so," I answered. "I sure hope so." Jack and Howard had turned up about ten that morning, and Jack had hopped out of the boat, calling excitedly that Howard had the stuff. For two hours they had been fiddling around on the shoal. I had been afraid of going out there for fear I would step on a stick or two. On the other hand, I didn't want the place blown up without having plenty of warning; so every now and then I wandered down as far as the willow tree to call and ask them how they were getting along. Finally Jack came clumping back from the shoal along the catwalk toward me—a little hastily, by the sound. Half laughing with excitement, he said: "You sure now? . . . Sure you want to go ahead with this? The Indian's got that shoal wired up till it looks like a spiderweb."

"We want a dock," I said, "but I don't want the place blown to bits."

"Well, you're not asking me for guarantees, are you? Cause, Jesus, I don't know nothing about that stuff. Tell you the truth, I don't think Howard does either. But Christ, I don't know. I'm the cement man, remember? The Indian, he's supposed to be the powder monkey!"

"But he's still out there!" I said. "If he doesn't know what he's doing, what the hell is he staying out there for? He could get his head blown off!"

"Yeah," Jack agreed, "he could, but a job's a job, eh? Look at it that way. Besides, I think he'd kind o' like to see what happens. I'd kind o' like to see myself. Wouldn't you?"

"I'd like it better on somebody else's property," I admitted.

"Well, this is yours, not mine," Jack laughed. "So what the hell, eh?" and he laughed again. "Hey, Indian," he shouted, "how you coming?"

"Pretty good," came Howard's answer from fifty feet out.

"How about the license?" I asked nervously. "Did he get the license?"

"Augh! Who the hell's gonna give him a license when he don't know a stick of dynamite from his aunt's elbow?" He giggled happily.

"Don't you think you ought to go see what he's up to?" I asked.

"Go to hell!" he laughed. "He's getting too near the end for me." He took two steps back and I backed with him. "Got nerve, though. You got to give him that. Or maybe he's just dumber than us, eh? Wouldn't catch me out there. But I would sort o' like to watch."

"Just about ready," Howard called, "but I need a bag of sand to put on the powder to tamp her down."

I produced an old sack and we filled it with gravel from the shore. As Jack took it cautiously out to Howard, I called after him, "Hey, Jack! How about . . . Maybe we ought to do the cementing first and blast later . . . some other day, maybe?"

"What?" he threw back over his shoulder, "blast after my cement's down? Smash it to bits? We're just about set to go now, and anyways, Howard can't get the scow till Saturday."

"Remember!" I called back, "it's the *right* side, the *right* side of the shoal I want cleaned off. Make sure you've got the charge on *that* side."

"Listen to him," Jack shouted to Howard so that I would hear, "maybe we're dumb, but we aren't deaf, for Chrissake. Must be the twentieth time he's told us that." Their voices muttered to each other as they adjusted the sandbag.

"The *right* side as you face the river," I called.

"Augh, shut up, Russell," Jack shouted, "don't interfere with experts at work."

"There we are,"said Howard with satisfaction, and Jack came rattling back at top speed over the loose rocks. "Under cover, now. We're going to blow her!"

"This far enough?" I shouted from behind the willow tree as Jack dashed past me making for the house.

"Hell no," cried Howard. "Get up in the house. We might take that tree right out of there."

"God! You better not," I shouted back, stomach knotted, making for the porch in hot pursuit of Jack.

"Ready?" came Howard's voice from the distant rocks.

"Is he really going to?" asked Elisabeth, her hand trembling in mine or mine in hers or maybe both.

"I guess so," I answered.

"But what was that about the willow?"

"Three!" Howard called.

"Inside, Richie. You too, Mark." Elisabeth gasped. They didn't move.

"Two!" came the lonely voice from the shoal.

"O God!" Elisabeth murmured.

"One!" Electric silence.

"Fire!" and the house leaped under us.

"O good Jesus!" hollered Jack. "Look at that sandbag! It's going to go right through your neighbor's roof." Rocks thudded on the grass before us and a few landed on the roof, rolled off, and plunked down about us. Jack was running toward the catwalk.

"You all right, Howard?" he shouted.

"Hell yes," was the triumphant reply. "How was that one for size?"

"Come on," Jack yelled back, "let's see what she did."

I was rooted to the steps.

"What's the matter?" he called.

"Johnsons' roof!" I howled.

Jack stopped, looked up, and burst out laughing.

"Damn bag's just laying there. Course! It's *empty!* The blast knocked all the guts out of her, so it was only an empty bag when it landed."

We laughed with relief and sank onto the steps.

"Just an empty bag, Richie," I said. "Just an empty bag."

We went down to have a look at the shoal. Except for a smattering of rock bass and sunfish that had rained down after the shot, nothing seemed to have changed.

"She's cracked, though," said Howard, "cracked down under there. Come on, Jack, let's get the crowbars, and we'll pry off some of those loose points." I got a bar too, and the three of us hammered and pried and swore.

"Got some more powder," Howard said after a while, "so don't worry." I worried. "After smuggling her in, we're sure as hell gonna shoot it, eh?"

"Not unless it's really *necessary,"* I said. "We don't actually have to . . ."

"Oh yes, we actually do," was Jack's sharp reply. "I haven't had this much fun since I was a kid. How many chances do you get to be paid for something's much fun as this? We'll shoot it, all right, won't we, Howard?"

They did shoot it or, rather, Howard shot it while we cowered. After each blast, Richie and Mark hunted happily for fish among the rocks, while Jack, Howard, and I worked with the bars in our bathing suits, prying off the points and ledges of the huge outcropping to make a straight edge for docking. By the end of the day, Howard had shot all his dynamite and we had almost the last twenty-five feet of the outcropping clean to a depth of five feet.

"Just like she was sliced off with a bread knife," said Jack happily over his cold beer, "and we'll lay a six-foot strip of concrete all the way along the edge there. Why, you'll have the best goddamn dock on the whole river. Ain't no ice gonna take *that* out, by Jesus!"

On Saturday morning Jack and the Indian turned up, towing the scow laden with twenty bags of cement and a mixer. The scow was nothing more than a big shallow box, ten by fifteen by two feet with a hatch or trap door in the lid—"for bailing her," as Howard explained, "cos she leaks a bit."

"I didn't know a scow could leak," I said.

"This one can," he chuckled.

After unloading, Jack said: "We'll be back this afternoon with the gravel and we'll be all set." But they didn't come back.

Duffy blew in about noon Sunday.

"You got a pump?" he asked. "Them crazy buggers sunk the scow yesterday."

"Didn't know that was possible!" I exclaimed.

"Well, it isn't easy," he admitted. "Actually, they didn't so much sink it. They capsized it, and that's even harder."

"But how?"

"They overloaded her with gravel. They swung in too close to Treemont, and the wake of a tour boat just tipped her right over on a shoal near shore. Then they tied her to a tree and yelled like hell till somebody picked 'em up and took 'em back to town. They're out there sittin' on her now, and she's full of water. You got a pump?"

"Not one big enough to bail out that thing."

"How about beer?" I had a case. "Then get it and come on."

Duffy dropped me off on the all-but-submerged scow with the beer and went off to scrounge a pump.

"Christ, Russell!" complained Jack, opening three beers, "I thought you'd never get here. What the hell you think you're doin' lettin' me and this here poor Indian die of thirst?"

In half an hour Duffy was back with a little gasoline engine hooked up to a pump. Soon it was putt-putting away beside us and sending a steady stream from the hatch back into the river.

"This is going to take a while," Jack said philosophically, "so relax. Not a bad way to spend a Sunday afternoon anyways." It wasn't.

As the scow lifted slowly from the rocks, Jack and Howard carelessly threw aboard a shovel or two of gravel from the pile in the shallow water beside us, and by late afternoon we were afloat with a reasonable load on.

The cementing went smoothly. By Tuesday afternoon Jack, the Indian, and I sat on the catwalk with my twenty-five-foot strip of cement dock beginning to cure.

"That's going to be really something," I said with satisfaction.

"She'll be there longer'n you or me," Jack said contentedly. "Now you can tell your neighbor there that he can tie up at your place . . . well, tomorrow. Don't you dare go walking on my cement tonight. Throw a few pails of water on her to help slow down the curing, but don't go walking on it till tomorrow."

"What do you mean? It's not *your* cement."

"I put her down, didn't I?"

"Yeah, but I'm paying you," I argued.

"If you'd pay that fifteen-hundred they were talking about at Mitchell & Wilson, it'd still be mine because *I* put her down!"

He was right. The aching muscles and kink in the back constitute ownership beyond money.

"Right off the end, though, Jack, there's that big point sticking out just under the water. It would take the bottom out of any boat that pulled up there."

"Yeah, but nobody's gonna dock across the end, not when he can come in alongside. There's too much weather at the end."

"They ought to," I admitted, "but it seems a shame not to be able to use that end."

"Holy Christ, Indian! Let's get out of here. Cement's not dry yet and Russell's worrying about an extension!"

"Not an extension," I said. "Just the end! If you could just blast off that big point . . ."

He put down his empty can and he and Howard climbed into their boat. As he left, he shouted: "If you go blasting anywhere near my cement, I'll have the hide right off'n you!"

Before breakfast next morning I said: "Come on. Let's move the boat over to *our* dock."

"You think it's really all right?" Elisabeth asked.

As we walked to the dock we met Dick Johnson.

"Good morning," said Elisabeth; "we're just going to move the boat."

"It looks great, just great," Dick said.

"We're sure glad to get out of your hair at last, Dick, but I do want to thank you for your kindness and hospitality all this time."

"That's all right," he said thoughtfully. "To tell the truth . . . I know this is a great day for you and you must be very happy, but, to tell the truth, I'm just a little sad you're leaving."

"We're not moving away, Dick," I laughed. "No danger."

"No," he said. "No, but . . . funny . . . I'm glad for you but . . . well, you've just been here, that's all, and it just seems a little sad."

"I'll come over and borrow your drill after breakfast," I consoled.

"You do that," he laughed. "Seventy-five feet isn't all that far, I guess." Under my own happiness, I recognized his sadness. The pattern of our summer lives would be changing, and every change has its darker side.

We untied and pushed off, putting the few feet to our dock. As I made the bowline fast to the mooring ring that had been sunk in the cement, the roughness scraped my knuckles with the abrasiveness of new things. It would take time, time. Time would smooth it down, as it taught me how to handle this new fact.

"Do you think it's really safe to walk on?" Elisabeth asked.

"If we can't walk there, we can't walk anywhere," I said, climbing up. I stamped on the cement till my feet stung.

"Be careful!" she cautioned, getting out gingerly.

"Come on," I said. "Let's dance," and the wind blew fresh and sweet upon us. "I'd never be a decent Marxist," I crowed.

"Neither would I," she agreed. Then, reflectively, "But sometimes, you know . . . sometimes I think we're crazy. While all those other people are taking vacations, we're up here painting and hammering, getting slivers, and springing up in the middle of the night when it storms to march through the rain to check on the boat. . . . And all those other people, they're sitting on the beach at Atlantic City or somewhere getting tans. What do we do it all for? We're just bloody maintenance men, that's all—bloody maintenance men!

"Mm," she continued, as we went along the fifty feet of ancient catwalk from the new cement to the willow tree and the shore, "this catwalk seems kind of wonkie, don't you think?"

"Sure is," I agreed, "and there's the real point."

"Where?" she demanded.

"Well, we just can't help it."

"You mean, it can't be fixed?"

"No. What I mean is that we can't stop getting slivers. We've just tied up the boat back there for the first time, and you point out what needs to be done next. It's just part of our nature, that's all."

"It's your nature," she said dryly, sitting down on the top step. "I don't care what you say! Other people live sensible lives, but not we!"

"You mean you'd rather have a new car?" I asked, sitting down beside her.

"Oh no," she said contemptuously. "I couldn't care less about a car, but I'd like a new dress once in a while."

"Then get one," I urged. "Get two."

"I can't, and you know I can't! Those are my dresses out there on the shoal . . . all cement."

"Like my shoes," I said, waving my tattered canvas loafer, whose sole flapped mournfully. "But what do you remember about your childhood?"

"What do you mean?"

"I mean, what stands out as being wonderful back then?"

"Orkney, the sea, walks along the cliffs, and the river in Cornwall, the shine of the mudflats . . . it was all so beautiful."

"Well, what do you think our kids will remember? You want them to remember the scruffy patch of grass between the semi-detached houses in Lancaster? Trimmed hedges, the roar of power mowers, and steaming nights when nobody can sleep? Or would you rather have them remember breakfast on the back porch, having tea here on the front steps looking through the willow tree out onto the Forty Acres, the wind, jumping off that great cement dock into sixteen feet of sweet fresh water whenever it got hot, reading Tolkien in front of the fire in the evenings? What can we actually give our children except things to remember? And if that's all, why shouldn't those things be beautiful?"

"Oh, shut up," she said, half annoyed, half moved, "you're just saying that because you know I can't resist the idea of something for the children. You can't fool me," and she sniffed. "We're not doing all this for the children."

"No," I confessed, "not entirely . . . not altogether. But anyway, what I said is still true, isn't it? What else can we actually give them? Money? We haven't got any. Even if we did, they'd only spend it, and it would be gone. Memories, you can't spend. They're part of you, just like your nose, but they're the best part."

"You really think they'll bring *their* children back here?"

"Course they will. Just wait! These islands are full of families that have been coming for generations. Once you get bitten by this river, there isn't any other place, that's all."

"I wish ours could see Orkney and Cornwall," she said wistfully.

"Sure. That's because you love those places, and they will see them. They will, but for them this is their Cornwall and Orkney."

Bare feet thudded in the house behind us and the door opened.

"Hey, Mom," Richie asked, "what's for breakfast?"

"Save the children!" I cried, getting up. "The poor things are starving!" She laughed as she rose.

"This top step is rotting, you know, Bob," she said as we went in, "and I bet the joists, or whatever you call them, are going too."

"Probably are," I agreed, "but first I'm going to knock off that point of rock out there at the end. Then a boat can tie up there too." And we went to the kitchen.

"How big is it?" asked a geologist friend who summered on Eagle's Crag.

"It's about eighteen inches wide, sticks out about three feet, and is maybe two and a half feet thick or deep. Sticks right out from the end of the dock, and it would knock a hole in any boat that pulled up there. How can I get rid of it?"

"Stick of dynamite would do the job," he said. I remembered Jack's warning as he had departed.

"But my cement topping. A blast might crack that."

"Could," he acknowledged. "In fact, probably would, but there's only one way to find out, and you said you wanted to get rid of that point."

"Yeah, I do, but I don't dare blast."

"Then how about chewing it off?" he laughed. "Hey, wasn't that a wind last night?" and the conversation drifted off, leaving me with my problem. Everyone I asked offered the same solution: Blast. It wasn't much use.

"That's how I got married," said Norman Zacour, a medievalist who was renting one of the little cottages on the other side of the Johnsons, one of the two we almost bought and were now very glad we hadn't. He was teaching summer school at Queens University in Kingston and buzzed off to the mainland every morning to drive the twenty miles for his lectures.

There was a break underground in my soil pipe that led to the cesspool and he was helping me dig it up.

"What do you mean, that's how you got married?"

"Toilets," he explained. "Toilets. I had this blind date with Betty in Winnepeg. I took her home, and when I got back to my apartment, there she was on the phone telling me that her toilet was overflowing and what could she do about it?" He drove his shovel into the clay. "There's romance for you."

"But that doesn't explain how you got married," I objected.

"Well, that's because you don't listen, goddamn it! I went over and fixed it, and that was that! Come on! I think we're getting there. I just hit the pipe."

He had indeed struck it rich, and after a little more digging, we dragged out the broken section of pipe and set off for town to get a new piece. On the way in, I broached my favorite subject: "Hey, Norman! You're a historian, a medievalist . . ."

"So?" he said suspiciously.

"Well, how did they split rock in those days?"

"Rock? Split rock? What the hell would they want to do anything like that for? Ask me something about Bernard of Clairvaux or the papacy or something . . . something really interesting. Anyway, what do you care how they split rock . . . if they ever did?"

I explained my problem, closing with:

"And they couldn't have had dynamite then, but they split rock, all right. They had to split rock to build all those churches and cathedrals!"

"They built a lot of those," he admitted. "It's not generally known, but they were all prefabs. Not a split rock in the bunch! . . . You're supposed to look to the Middle Ages for things like wisdom and so on—not for silly things like splitting rock. Anyway, there must be somebody right here in Gananoque who knows how. We'll ask them."

"You think maybe we just ought to go up and down the street asking? . . ."

"No!" he said. "Look! I know it takes longer up here in the fresh air and everything, but even here people die. They put them in cemeteries, and then they put rocks on their heads—tombstones, they call them. For reasons that escape me at the moment, it's important for those tombstones to be square or squar*ish,* and since they don't just grow that shape, somebody's got to split them; so we just find the guy that makes tombstones and we ask him. It's perfectly simple."

So we did that perfectly simple thing.

His name was Mr. Williams, and his establishment was a shed almost as ancient as he was.

"Sure," he quavered. "We used to do that, but everything's done by machine nowadays. Not like the old times when we had to use plug and feathers."

I asked him what plug and feathers meant or were. He tried to explain, then gave up. "Maybe I still got some around here," and he tottered over to a shelf in the corner where he rummaged among some cans. "Here we are," he said, tottering back and handing me a can. "Take 'em along with you. I've got no more use for them."

"Thanks," I said, "thanks very much, but I don't know what to do with them."

"What you do," he explained, "is you drill a hole, many as you need . . ."

"Drill a hole in what?" I interrupted.

"Why, the rock!" he said, a tinge of annoyance in his voice, "didn't you say you wanted to? . . ."

"Yes, yes, I do," I said eagerly, "but I didn't know . . . you see, I mean, how do I drill a hole in a rock?"

"Why, with a star drill!" he said, as if everyone knew that. He was having trouble fathoming the real depth of my ignorance despite my best efforts to expose it. I told him I didn't know what a star drill was.

"A star drill's about like a ten-inch spike, only instead of a point, it's got kind of a cross at the end, and you just pound that into the rock. Pound a few times, turn it maybe a quarter of a turn, pound again, and so on. Like I say, you drill these holes . . ."

"How far apart must they be and how deep?"

"About three inches deep and maybe six or eight inches apart. When you get your holes, you slip in two of these feathers." He handed me a small piece of steel. It was a tapered strip of half-round, bending back at the thin end. "You slide two feathers into the hole, flat sides together with these thin tips spreading apart out the top. Then you slip one of these plugs like this"—he gave me a small chisel—"in between the feathers. Do this in each hole, and then just go along tapping these here chisels—plugs, we call 'em. Don't have to hit 'em hard, either. Just drive each one in same as the others."

"And then what?" I asked, because he seemed to think he had finished.

"That's all," he said with surprise. "The rock splits."

"You mean it really does?"

"Sure it does," Norman assumed the explanation with exaggerated patience. "You see, what happens is that, as you drive in the plugs, they put a line of tremendous outward pressure on the rock through these tapered feathers. The farther the plug goes in, the greater the pressure, until bang! she splits! That's how they used to split rock in the Middle Ages—you know, for all those churches and cathedrals. It's not generally known, but that's how they did it!"

"He's right," said Mr. Williams. "Just make sure the flat sides of your feathers in the different holes are lined up with each other so that the pressure from the plugs all goes in the same direction. And crack! Just like your friend there says."

"I guess I understand, Mr. Williams. Thank you. May I borrow these plug and feathers? I've got this rock, you see. . . ."

"Take 'em," he said firmly. "I don't want 'em no more, now that I got electric. You got electric out there?" he inquired. I said I had. "Be lots quicker, then, to use an electric drill with a carbeloid bit. It's the drilling that takes the time."

"But I've got to do the drilling under water," I said.

"Oh," he said, "then you can't hardly use electric, can you? You can drill, all right, with the star drill. Takes time, that's all, and you got to keep at it."

"Right!" agreed Norman. "Thanks a lot, Mr. Williams. We've got to pick up a star drill. Bob's got a lot of drilling to do. I'm a toilet man, myself."

"Where can we get a star drill?" I asked.

"Hardware store," he answered. "Good luck. Remember, you got to keep at it. . . ."

He was right—about "keeping at it," that is. In my bathing suit, I straddled the point of rock, my back to the waves, eighty-eight-cent star drill held firmly against the rock in my left hand and the hammer in my right. Whack, whack! turn the drill. Whack, whack! turn. Whack, whack! turn.

The drill actually did go in, but it took three days to drill three holes. I could see why Mr. Williams, however much he mourned the passing of the old times, didn't want his plug and feathers back. At last the time came to drop the feathers in the holes as he had instructed and insert the plugs between them.

"Don't you split off that rock before I get back from Kingston," Norman warned. "Goddamn it, I've got to be here!"

Next afternoon, crouching on the cement above me, munching his sandwich, he said: "Okay. Now you have my permission."

I began tapping in the plugs, three taps per chisel. Gradually they went down between the feathers. As I tapped number one, he reached down with a hammer and hit number two.

"Hey!" I complained, "leave that alone. This is my rock."

"Jesus!" he cried, leaning over, "there's a crack . . . actually a crack. I think you're cracking it!" Tap, tap . . . a wave slapped my back as I raised the hammer for the third, but my perch quivered and I squeezed it with my knees to hold it still. Then I was gulping water and struggling frantically. The rock had broken and dragged me down before I could snatch a breath. There was the heavy thump of rock on rock at the bottom, and I fought my way up to Norman's laughter.

"Jesus! It's really gone! Look at that . . . just look at *that!* As if you had sliced it off with a carving knife." He took my hand and rubbed it down a smooth face of rock.

"Augh, come on," I spluttered.

"No! Really. I'm not kidding. That's where it was. Why, the *Queen Mary* could pull right in here now!"

Climbing up, I sat down on the cement to try to digest the marvel.

Ever since the glacier had ground out this huge riverbed and withdrawn, that point on which I had sat had stuck out from this big outcropping. How many thousands—or was it millions—of years had it stuck out? With a hammer and star drill and plug and feathers, I had carved it off. It had plunged to the bottom where it would lie for eons. I had changed the face of the world. Not much, but because I had been born, that chunk of rock had been split, cut off. I had made a mark in something more permanent than the sands of time. That mark would remain longer than the greatest of Shakespeare's plays. Maybe only four or five people would know I had knocked off that point, but I would know for a little while—for as long as I knew anything. The very rock on which the earth rested had been changed because of my passing. Perhaps. . . . No, not perhaps. I actually must be something more than a fleck of time-tossed foam. There must be substance in me. I had cracked the earth. If I did no more, I had done something.

What had I done in my ten years of teaching to compare with this? Influenced students? Some, perhaps. A few very kindly had said that I had. And maybe they spoke the truth. After all, my own life had been changed by instructors who had shown me the value of literature, but did they know that they had? I had written to several of them to try to explain what they had meant to me, but they had only my words, words that must fade, be misplaced, or forgotten. This smooth scar on the face of the shoal would not forget. It was a palpable fact, not a word, not a pattern of waves in the air to be blown away on the wind or a set of chicken scratches on a flimsy bit of paper to be shoved into a drawer, finally to be crumpled up and thrown away. Those instructors, like me, must have their moments of doubting. Was all that talking actually worth anything? Anything at all? What effect did any of it have, and on whom? And did those whom it did effect, did they even know themselves what that effect was? The rock had acknowledged my hammer in clear and irrevocable terms, and for eternity. What I felt about that must be what people called satisfaction. I know I could not hope for it often, so it was all the more to be treasured.

"Elisabeth has to see this," I said.

She came; she saw; she admired.

"Someone could really tie up across there now, couldn't they?" she congratulated. "Who would have believed it? They couldn't stay very long against that rock, but they really could come in there."

"It's going to be better when I finish," I said.

"Better? What do you mean?"

"Well, why couldn't I attach a long plank, maybe twenty feet, all along that rock face?" Norman wanted to know how. "Star drill," I explained. "I could drill holes in the rock, put in those lead sleeves that expand when you screw a lag-bolt into them, and bolt a plank right across that whole front end so that a boat wouldn't even touch the shoal."

"You could build a whole goddamn dock across the front that way," he agreed. "Come on. Where's the star drill?"

And we did, and topped it off with a six-foot diving platform.

6

Ben's Story

One afternoon in our second summer, when our dock was still a dream, Elisabeth and I were relaxing on the front porch with cups of tea and considering our general good fortune—despite the fact that in spots the boards of the porch had a distinct bounce, strong evidence that all was not well with the joists beneath. But everything takes time, I reflected. Besides, if one thought about the *back* porch—which one didn't, but which Elisabeth did—that gave much more cause for concern. The back porch had not enough spring even to bounce. It merely sagged. Also, large black ants seemed to be attending to the underpinnings vigorously. Let them. Out front the sun shone, the breeze was fresh, and off to the left a catbird sang of summer. An occasional splash and shout from the river's edge proved that Richie and Mark were happy in their world. In her rush-bottomed rocker, a full teapot beside her, Elisabeth had the blue and silver of the river before her. On the top step, digging aimlessly with a fingernail at a spongy board, I let the afternoon flow over, around, and through me.

With a sudden cry Elisabeth jumped up, ran to the side porch, and vaulted the railing. She made for Johnsons' dock at full speed.

"What's the matter?" I cried, starting after her, but she made no answer. A few moments later she panted back with a dripping, spluttering Mark in her arms.

"Is he all right?"

"Yes . . . yes. He's all right. I just happened to look over, and there was this thing floating in the water beside the dock, and Markie wasn't on the dock any more!"

Mark was coughing and beginning to cry. I took him from her.

"God bless life jackets," I said.

"The water wasn't deep, but he was lying there with his face under. Just scared, too scared to do anything."

"That's a high dock," I said. "That's a long fall—enough to scare anybody! Lucky he didn't break his neck!"

Mark recovered in a few minutes, but we didn't.

"Life jackets!" I declared to Richie and Mark, *"life* jackets! Maybe now you'll understand. There absolutely *can't* be any going outside this house—not *anywhere*—unless you two are wearing, not carrying, but actually *wearing* life jackets! Understand?" They said they did. "Well, you'd *better* understand, because in this house that's the *law!"*

Of course, it had been the law, but they had dodged it whenever they could. Like all other laws, it seemed to carry with it its own invitation to be challenged or evaded. Their testing of its strength went something like this:

"Can I just open the front door, Dad?"

"Sure."

"How about standing with one foot outside on the porch?"

"I guess so."

"Two feet outside?"

"No."

"But who needs a life jacket on the porch? There's no water *there*. I'm even under the roof, so I'm not really out of the house."

"All right, but just on the porch."

"How about the steps? They're part of the porch, aren't they?"

"Suppose they are."

"How about my feet on the grass at the bottom?"

"No, damn it!"

"But why not? Grass isn't even damp! Why can't I? I'll just stay right up by the house."

"I said NO!"

"But why?"

"Because you won't stay there, that's why. Now, get back inside this house and don't you put so much as a foot outside—even onto the porch—without a life jacket, do you *hear?*"

All children are practicing lawyers, or maybe it's the other way around.

This law was too important to be continually tested. Still, it was hard to remember every time a door opened to shout: "Hey, have you got your life jacket on?" and then to actually check. And sometimes we would wake to a silent house. Why silent? Were they asleep or already outside? The only solution for this problem was religiously to lock the doors every night—a precaution we never took on the mainland. That's when I discovered that we actually had six doors and keys for only two. For those other four I went to bolts, or, rather, climbed to bolts, installing them at the very top where the children couldn't reach them even from chairs. One morning I woke to hear Richie dragging the stepladder to the back door, so the stepladder had to be brought upstairs and stashed in a niche just outside our bedroom. Then we felt moderately secure.

Guests with small children walked unsuspectingly into the same problem. Almost invariably they came unequipped, unless we had thought to warn them. We were always having to scrounge for life jackets for visiting children, but it was well worth it. Richie and Mark needed the company, and small children were scarce. That's partly why we were happy whenever Ben took over his father's cottage down the island with his four. His love affair with the river stretched back for more than thirty years, as far back as he could remember, and his only complaint was that he had no cottage of his own.

"Someday," he said one evening as he was helping me cap a sixty-four-quart batch of home-brew, "someday I'm going to get out of that rat race in Toronto and live like a rich American, Russell." Ben was a molder in his father's hat business in Toronto, and he was a genius with his hands. "All I need is for someone to give me a lot, that's all. Doesn't have to be a big one either. A small one would do. I'm not greedy. You've got a nice piece of frontage here. Just a small one—say maybe down there in the kind of marsh? I'm collecting my materials. Have been for years." He passed me a fresh ginger-ale bottle full of nectar to cap and switched the siphon to the next. "Every Saturday I'm up at six, and me and my chum check the junkyards. The garage is so full now my car has to sit out front, and the cellar's coming on. Think this brew is ready? Here's the last one. How long do you have to let this stuff age? That first one must be at least an hour old. How about trying it?" We did.

"You've been holding back on me," he complained. "You haven't given me one decent problem to work on since I walked in the door. Getting polite or something? My rates are high, but you've got to admit I do good work."

I went to the top shelf in the corner and fished around till I found it.

"Well, there's this crazy anchor rope affair that's supposed to clamp the line as soon as the anchor hits bottom." I presented him with the small device with its pulley and spring clamp that wouldn't work.

"That's more like it," he said cheerfully. "You see, my problem is, I just don't have enough problems—not like this, I mean. Jesus! What kind of a Chinaman made this contraption? Whoever he was, he drank, because the way this thing's hooked up, it can't possibly work. No kidding. I need a place of my own. My father's too smart for me. He does all the interesting stuff around the place and leaves the bonehead jobs for me. Anyway, Pat's always nervous down there. It's not *ours,* and these kids . . . well, you know what kids are. They break a cup or something, and it's somebody else's cup. That's no good. Don't get me wrong! My father wouldn't complain if they broke the whole set, but you know how it is. It's not *ours!* That's what it comes down to.

"And two weeks! What the hell can you do on this river in only two weeks!"

"How about your house in Toronto? Can't you work on that?"

"Yeah. Yeah, I do, but it's not the same, you know. The kids . . . what can they do? Go out in the street and just play stickball. What kind of life is that? When I was Eddie's age [Eddie was five and Richie's chum] I used to take our old skiff and row to town for a loaf of bread. Well, I didn't actually row—not if I could help it. I'd get out there in the channel and wait for a motorboat—there weren't so many in those days—and thumb a tow to Gan. But Eddie! Geez, Eddie can't even swim a stroke. I can't remember when I couldn't swim. Matter of fact, I fished him out yesterday. Crazy kid had walked straight in! Yeah, he actually walked straight in, and it drops off to beat hell down there, you know. Scared the daylights out of Pat, but he wanted to swim. What kid doesn't want to swim? I don't want him being scared of the river. Why, this river is my whole childhood, for God's sake! That poor kid has nothing but a crummy old Toronto street."

"Did he have a life jacket on?"

"That's the rule, you know, but no, he didn't have one on. Christ, they can't sleep in them. I didn't, and he's no different. Can you keep life jackets on yours? I mean *all* the time?"

"We try."

"Yeah, but can you do it? That's what I want to know. And you know, just for two weeks and maybe the odd weekend . . . they don't get used to wearing the damned things. You wait. You just wait! I'm going to build my own place. Actually, Pat doesn't know yet, but I've got my fifty feet. Settled the deal yesterday."

"Hey, Elisabeth!" I cried, "Ben's got a lot . . . he's bought a lot!"

She rushed out and we both congratulated him.

"Let's open the second one. A celebration is in order. There's nobody we'd rather have for neighbors, and the kids will have company!"

"Going to come down weekends this fall and put in the foundations. Going to start with a prefab garage and move on from there, French Provincial, I think they call it. Anyway, it'll be someplace to get in out of the rain. That's all you really need to start with, and in a couple of years I'm going to have things just the way I want them. Then the kids can break *our* cups! No apologies to anybody then."

He was as good as his word. That September he dug the footings and put in the foundations of his dream. Then he arranged to rent a trailer to haul down his prefab. As they often did, Ben's parents came down to the island early on Friday for the weekend, and they brought Eddie to supervise Ben's construction on the morrow. In the confusion of arrival and getting settled, Eddie slipped out of sight. Missing him, the grandparents went looking. They found his body floating in the slip of their boathouse, fifty feet from their front door.

The following summer the island lay dark under the shadow of Eddie's death, and only slowly, very slowly, did it soften. Far up the shore at our house, it lay in the silences between speech. It was not always there, but when it came, it seemed to say: "I have not been far, and when I leave, I shall not go far because I am now and shall be part of this place and a part of your feelings for it."

And it was so.

The black center, the burden of zero, lay elsewhere, lay in the hearts of Eddie's family. Though we were only under its edge, it touched, enclosed us, grouped us with those who suffered most. Eddie's going filtered the sun as it blessed our breakfasts on the back porch, our afternoons in front, and muted the hiss and slap of waters. And over the door of that new way of life that we seemed almost driven to build, his going stamped a sign in letters so bright, so large that even the blind could not mistake them. It read: TRANSIENTS!

Gradually we had understood that our coming to the island had been not so much a coming to, but a fleeing from. Deeply we had hoped to escape from the false and trivial pressures of routine life on the mainland, with all that mainland meant. And we had escaped from many; but more than we had supposed possible, we had begun to feel the joy of confronting and living with realities. So exhilarating was this sense that we discovered a latent power to build, and it seemed necessary that what we built must be for eternity, something that would outlast the sullen furies of winter storm and summer lightning. The dock we would build, for example, would not sag, crack, or wash away—not till bedrock itself should shudder, move, and melt. In our thoughtless joy we believed that it was possible to do this, that we were not transients, that somehow we must be permanent. On this illusion the shadow lay its thin, small hand, and nothing would again be as it had been. We had been in the presence of too much loveliness, and Eddie's passing was the tightening of the throat muscles, a reminder that, after all, we are very weak, very small creatures, and we must not expect for ourselves and for the earth more than is possible. His passing brought us news of our own going. Many things we could change; some, we would; but not this one. This news bore the seal of mortality.

We mourned slowly and long. We grieved for the boy, for all that now he would not do, feel, and be; and for all the joy, comfort, and pain he would not give his family—banging down the empty milk glass, pushing back the kitchen chair, and slamming the door too hard behind him as he ran out to play or went off to college. They would never have to get him away from the TV and pack him off to bed; never have to break his abstraction at the dinner table to pass the salt. They would never have to yell at him to pick up the boots he had left in the middle of the kitchen floor after stamping in, chilled to the bone, on winter mornings back from his paper route. Not ever would they have to do those things—not ever—and ever is a very long time.

What we felt had been but the breeze from the blow that had narrowly passed us. Theirs had been the deepest wound. What would it mean to their lives whose fabric had been so torn? Could they recover? Could they demonstrate the greatest and simplest proof of life, the capacity to heal? It would take great courage to accept the workings of that capacity. When the wound is too deep, the tissues have undergone such pain as not to be willing to be joined again; they may refuse. Then that which was living is merely clay. If they have the courage to consent, the shattered bone will be, in time, stronger for having received the blow that broke it. More tightly, more fiercely do the tissues knot, tie, twine, mingle, tangle, clench to one another—more tightly even than at first, because now they know the force of that which they must resist.

"Pat will never come again," Ben said hazily on our steps one afternoon late in the following summer. "This is the first time I've been back myself, and I sit here wondering why I came."

"Could you help coming?" I asked.

"No. Not really. Not really," he sighed as Richie and Markie squabbled inside as they pretended to water-ski across the dining-room floor with cigar boxes for skis. The river rolled and slapped at the rocks before us. "I can't hate it, though," he almost apologized; "I just can't. She can. She can hate everything and everybody, and I don't blame her. I can understand that. I don't blame her. I mean, what do you do when your life blows up? It doesn't blow up, really, because it's easy to live with an explosion. An explosion's over right away. This doesn't end."

"I don't know what you do, Ben, but you can't unlive all those years here."

"I don't know," he said thoughtfully. "Maybe I should . . . sometimes I think I should, but I don't know if I'm able. I know how she feels . . . no, I don't know. I don't even know how I feel, but I know I can't stay away. I stayed all year, but this morning I just couldn't. I got up and told her I was going. Didn't even say . . . I didn't have to say where. She knew."

"What did she say?" I asked.

"Nothing. She didn't say anything," he sighed. "But, Bob, I left her there. Just left her. I didn't want to. I don't want to even now, but I did, I have, and I can't help it. Something's wrong with me, I guess."

"Something's wrong with all of us," I said.

"But you wouldn't do *that,* would you? You wouldn't leave Liz like that. . . ."

"I don't know what I'd do, Ben."

"She needs me. God knows I know that, but I went and left her to come back here and look . . . well, at the river, my foundations . . . the whole ball of wax. You know, I just sat there and stared at those foundations." He thought for a while. "But what I want to know is how do you give consolation to somebody? How do you give it when you need it yourself?" I couldn't answer. "You know," he went on, "when I first saw the river . . . again, I mean . . . I was wondering all the way down how I was going to feel about it, but when I pulled up at the Bay Street Dock, the river was just there, that's all, just there like it always had been, like nothing had happened. And, Bob, I wasn't even mad. I didn't know before I got there whether I could get out of the car, but I did. I guess I'm still kind of in a daze. The river's all right. It's me that's confused. It's sort of like I got to relearn everything. Does that make any sense?"

"I think so," I said.

And I felt the same loneliness in the wind or behind the wind that I had remembered when it first blew over me. When had it first blown? I had known that lovely loneliness of the empty wind for years—or was it centuries before I was born?—and I would know it in the centuries to come. It would blow over the grass, and I, somewhere, would hear and feel it, the gentle messenger of death with soft news from the quiet of eternity.

Time's stage is too big for the comfort of us mortal players. Why should we not huddle together and sing to keep one another company? The songs of our tongues, our hammers, and our dreams lend substance to our fragilities, weaving for us a common web of life to stand on, to build from.

"It makes sense, Ben," I said, "and I bet . . . what's more, I really bet that someday you'll actually build on those foundations. Maybe not because you want to, but because you can't *not* build."

He was silent. "This place took my son, but here I am. I came back. Maybe I will, but I don't understand that . . . not now, I don't. Do you?"

"I think so," I answered.

"Then, will you tell *me?* What is there about this place that does such things to us? Oh, it's beautiful, all right, but so are lots of places. I'm not trying to be smart or anything . . . excuse me if I'm talking out of turn, but for you . . . well, it's not beautiful for you—not like it is for the rest of us. Bob, all the time it is beautiful. No matter what the weather, it *is!* Sounds crazy, but even on a gray day when the wind's up . . . I mean . . . for example, I come down here duck hunting! Never shot a duck in my life, and I'm not going to. Cold? Jesus it was cold, but it was lovely. Well, you don't see all that! Bob, I've actually got up here in the middle of the night, gone out in the boat and just sat . . . just sat and looked at the stars and their reflections in the water. Well, maybe *I* would come back here to do that, but you wouldn't. Why do *you* come? I'm serious because, you see, the reason that *you* come is probably the real reason why I can't stay away, because I don't come back only for the beauty. It's something else, and that something else is probably what brings you. So what is it? Do you know what it is?"

"Sorry," I said. "It's there, Ben, but I don't know how to say it. Maybe it's that this is the only place I've found where I can believe in death, so I wouldn't mind dying here. Anywhere else, dying would seem like a cheat. That sounds crazy but . . . oh hell, I don't know."

He sat in silence for a long while, then got up. "Well, whatever it is, I've got to be moving. Nice talking to you." We shook hands. "Be seeing you . . . maybe . . . probably, but I don't know when."

"Give our love to Pat," I said. "We'd be happy to see her here."

"If you do," he said, "it'll be years. So long," and he left.

Pat's heart was clenched against the river and life on the river. For her, the summer nights with their starshine, moon-silvered grass by silent trees held no special mystery. They did not call forth her spirit for easy wanderings. Why should she go back to that exact spot, get out of a boat, and perhaps step on the very plank from which Eddie had slipped? If Ben had to have a place like that—if he could not live without one—why must it be there? Were there not hundreds, thousands of lakes, lakes neither of them knew, had never even heard of? If his need was so great, why could they not find one of these where he could work with his hands, saw lumber and hammer nails, and build a place that would be theirs?

It would be new for them both—fresh, clean, untainted, something they could start together, somewhere she would not have a handicap of his forty years of memories to contend with. Why could they not do that? Why not at least try?

Though tied to it, Ben, too, had been wounded by the river. Also, he was deeply tied to Pat; so he did try. For years he tried.

Like a restless ghost, weekend after weekend, he wandered from lake to lake, trying to look, trying to see the land and water that lay about him, but the eyes through which he tried to look were half glazed, the lenses clouded with too many memories. He came—not often—but he did come back to the river during those years of searching. Not so much "came," though, as fled. Breathless he seemed, almost as if he were surfacing after a dive so deep as nearly to burst his lungs. He would stay only a day, sometimes only an afternoon—just long enough to snatch a few gulps of air, just enough to let the pounding heart return to a throb; then down again.

One afternoon in the third or fourth summer of his quest, he sank wearily onto our front steps. After a few minutes of trivia, he lapsed into silence; then broke with: "I'm tired . . . plain beat, if the truth be told." Embarrassed, I muttered something about getting to bed at a reasonable hour. "Oh . . ." he brushed my remark aside and sat quiet for a while, "I can't do it. I give up! I'm beat . . . just beat, and she knows it. I won't ask her to come back . . . not here . . . but I can't stay away . . . not any longer, I can't. You understand?" I said I thought so. "Good, because I'm not much of a hand at explanations. If it comes to that, I can't explain it even to myself. I've looked, you know, but they're not the same. The water in those lakes doesn't *go* anywhere. It just lays there. But *here!* Why, you could shove off and go . . . well, to Tahiti, Australia . . . anywhere. Anywhere. Oh, I know what you're going to say, and it's true. Christ! I don't go anywhere . . . never did. Least not much farther than over to town for a beer at the Legion. Maybe once or twice I've gone as far as down to the Bridge, and what's that? Maybe ten, twelve miles at the outside. But the point is that I could . . . you could if you wanted to from here. Those other places . . . I don't know . . . I feel all boxed in like. Silly, maybe, but that's how I feel." He paused. "Not much of a reason, is it . . . not for coming back . . . for not being able to stay away." I said I thought it was a pretty good one, and that I knew what he meant, and I didn't go anywhere either.

"Didn't we talk about all this once before a long time ago . . . about what these islands do to people . . . how they hold them . . . how they hold us, really?"

"We tried," I agreed, "but I don't think we went very far then either."

"Well, didn't we just get finished admitting that neither of us went very far?"

I laughed and agreed.

"Well," he said, slapping me hard on the knee, his voice rising to good cheer, "what I'm saying . . . what I'm saying, geez, Russell, if you'd only listen . . . is that I've *decided.* Yes, by God, I've *decided.* I'm going to build! I'm going to build that damned cottage down there. By Christ, there's going to be no more of this laying around in bed snoring through Saturday mornings any more. I'm going to be hammering like hell down the way. Ben's palace is going up." I roared my congratulations and pumped his hand. "Got a backer and everything! Have I ever got a backer!" he repeated seriously. "Pat's been stashing away the odd dollar and not telling me anything about it. Well, the other night she turns to me and hands me this check and says, 'I hope this is enough to get you started.' And there I sat—just sat with my mouth open, and she said, 'Go ahead. Take it and get started.' Talk about a lucky man! I'm it, and I know it. Before I get finished, it's going to be the kind of place she'll want to come to. It's going to be *our* place!" He snapped back to joy. "Think I'll do her colonial style," he declared. "That'll set this river back on its ear. I can hear the tour boats slowing down with the loudspeaker explaining about 'that colonial structure—house?—no. Mansion, ladies and gentlemen! The colonial mansion on our right belongs to Mr. Ben . . .' "

After that, every Friday night he pulled in to the Bay Street Dock hauling a borrowed trailer packed with what, in anyone else's terms, would be junk lumber. By seven next morning his hammer was proclaiming to the sleepy world the joy of his heart that mounted to the shoulder, ran down the arm into the hand, and finally out through the wood and steel. Each series of blows began with one or two tentative taps, then crescendoed up to the final blow that declared, "And that's another one in, by God!"

As the frame slowly rose and took shape on the firm foundations, his spirits rose also. They were very high, indeed, when, one Friday, he did not get behind the wheel of an empty car. In the front seat beside him sat a quiet, bright-faced child eager with expectation.

"This is my Danny," he said proudly next day. "Remember him? He was only a baby, but here he is now. Danny, say hello. This is James." (By then there were two more children, James and Miranda.) Embarrassed hi's from the two six-year-olds. "Now look, you guys," he went on, binding them together in speech with ties that have grown, multiplied, and deepened ever since, "I want you both to look sharp . . . to pay attention to what us old men are doing around this island. Someday you two are going to have to repair what we're making, and it'll sure help then if you remember how it was done in the first place. Okay? All right, then. Now go play. Mr. Russell here was just going to get me a cold beer. Mind what I said about that life jacket, Danny," and the boys sidled bashfully off the porch while Ben sat down where he could keep an eye on them.

The next summer, with jacket, Danny regularly trotted up the shore by himself to hunt out James.

"He can stay, you know, Ben," Elisabeth said. "We'd be glad to have him. He'd be no trouble."

"Thanks," he replied. "I know there's nothing that Danny'd like better, and there's nothing I'd like better for him, but, Liz, I wouldn't dare go back without him. Oh, *I* would feel all right about him being here . . . I really would. I've come that far, but I couldn't face her—not without him. She's getting better, though. Maybe next summer. When he can go off that diving platform of yours down there without a life jacket, then he can stay."

Early next summer, on a glowing Saturday afternoon, he came trudging up with Danny and said, "Come on, everybody down to the dock. He's been practicing all winter at the Y, and he says he's ready to go."

"Just a second," muttered James, and he ran to put on his suit.

"Geez, Bob," Ben said quietly, "you ever been off that thing? Must be ten feet from the water." I admitted that it had been a long time, as we ambled toward the dock. "You afraid, Danny?"

"James can do it," he answered simply.

"But he's a Yank," Ben objected. "They're not all that bright. We've been over all that." Danny laughed. "It's not that I'd mind saving you," his father continued, "but I'm not going off that platform up there to do it."

James's bare feet thudded down the path behind us and onto the catwalk. We crossed the cement and sat down on the end beside the ladder that mounted to the platform. Casually James climbed up, marched to the end, and, without hesitation, dove.

"That's how you do it," he called back after surfacing. "Come on!" Without answering, Danny mounted the ladder. I could feel tensing beside me.

"Ready for a swim, Ben?" I joked. If he heard me, he didn't answer. Above and to our left, Danny's feet paused on the edge. Ben made a sound in his throat. At the splash, he was on his feet, poised either for a leap or a cheer. Danny came up, blew, and was swimming easily. Ben's hurrah bounced off Foresyth, a quarter of a mile in front, Eagle's Crag, Money Sunk, and faded out over the Forty Acres.

"All right, then," he murmured to himself; then, louder, "All right, boy, you *did* it! You earned your wings today. Now all you've got to do is talk Mrs. Russell here into letting you stay." Then, to me, "If you've got a beer, I could use one."

7

Guests

Guests. Why should any reasonable person object to them? Most people don't, but then, most people live on the mainland. In those nine months while I, too, am a resident, I sometimes wander about the living room halfheartedly kicking at the empty sofa and overstuffed chairs, asking myself why nobody ever drops in. But they don't. Up and down the silent street how many ears are half listening for the knock or chime that will not come to crack the glaze of TV monotony?

It seems a shame they don't come in the winter, because that's my time for being almost a responsible citizen. I keep my social outrage close at hand where it can be easily reached, hauled out, and flaunted as proof of my sterling qualities. I am ready, even eager, then to gloom over any one of the regular plights: the dying of the American city or its uncontrollable growth, the wisdom or folly of higher education as presently perpetrated, the tragedy of technocracy or its potential glory. Indeed, I stand ready with mustered eloquence to urge the adoption or abolition of any number of crucial or terrible things.

Then, sometime in March, the sun comes out, the geese pass over on their northward trek, and I begin waking to the voices of white-throated sparrows, robins, and thrushes as they consider whether this is far enough, or whether they might not better push on to the Adirondacks—say, Tupper Lake—with all those pines and hemlocks and larches. And from beneath, between, above, and through their thrusts of liquid melody they bring good news, that it is time for fresh business to be afoot in the world. Theirs is the gospel of the blood in motion, and under their sweet sermoning, my social, political, and professorial stances quiver, sag, shudder, and melt.

By late May, when I step from the water taxi onto the dock, I am neither a good nor useful citizen of any nation, no adoptionist or abolitionist for or against the policies of any group. It is enough for me to relearn how to taste the air I breathe, to feel the earth under my feet; enough to wonder whether any pipes, improperly drained, had burst over the winter; whether any trees were down; whether the shad were spawning up in the little Gananoque River because, if they were, the pike-fishing would be great for another few days; whether this summer I would actually get around to rebuilding the back porch. Private, personal, selfish, socially useless concerns, every one of them! Not a shred of worthiness, not a rag in the whole bunch to be admired by anyone. Yet, for me, they are enough.

A dreadful state, I suppose, for one to be in—so smug, self-centered, limited, satisfied, busy. Why, the thought of it is enough to make anyone not in that condition downright furious. No one should have the right to such . . . such . . . such happiness! This world is a blasted vale of tears and chock-full of suffering, injustice, and plain torment. What's more, anybody who doesn't understand that . . . anybody . . . anybody who doesn't face up to how bad things really are . . . well, he ought to be grabbed by the back of the neck and shaken till . . . till he bloody well believes it!

So these people, these others, come to visit. Probably I even deserve it, but I can't remember when, if ever, I enjoyed actually getting what I deserved. And the trouble is that they're nice people. They don't intend to be malicious. I even like them in an abstract sort of way. It's just that I don't like being reminded by the sound of their high heels and the swish of their sports jackets, reminded of what I ought to be but so assuredly am not.

There was Mrs. Blanchley-Hoyt, for example, the mother of one of Elisabeth's school friends back in Cornwall. The Blanchley-Hoyts, or rather, Mrs. Blanchley-Hoyt, had "been kind" to the Shaws in those old days—she felt certain that Elisabeth would remember—and she also felt sure that Elisabeth would be so happy to hear about all the worthwhile and important things Philip and Brenda, her dear old friends, were now doing; and how she, Mrs. Blanchley-Hoyt, was so eager to meet "dear Elisabeth's" children—she supposed they must be American children—and her husband—wasn't it Bob or something? Some other people to whom she had once been kind had invited her to visit Canada, and she would so love to come by, say from August fifth through the ninth, when she would have to move on to Montreal to see some other dear friends to whom she had also once been kind. She would bring her embroidery, be no trouble, and would so love to be with us all so she could return to her children, Brenda and Philip, with firsthand news of how Elisabeth was getting on.

"But I can't very well say no, can I?" Elisabeth observed.

"Why not?"

"Because I can't, that's all."

"You never told me you were all that fond of Brenda or Philip. They sound like the sort who would do pretty well in life. You sure you want to hear about that?"

"That's hardly the point, is it?"

"Seems like it to me," I said.

"Well, it isn't. I've got to say yes. Besides, she's not a bad old stick."

After hanging over us for months, August fifth descended, and our American children got a very full set of instructions from their mother on a great variety of subjects, ending with, "The main thing, the one thing to remember is to be polite, polite the *whole time,* and *no fighting!* Do you hear? NO FIGHTING! Is that clear?" They said it was.

"Well, Bob," said Mrs. Blanchley-Hoyt on the front porch after the flurry of arrival, "so this is your country house, is it?"

"No. Not exactly. Not our country house in the English sense."

"Oh no. Dear me, no. I can see that—not in the English sense, but I suppose it's your sort of version, is it?"

"Not . . ." I gulped, "well, this is our cottage, really."

"Ah! I see. I see!" she said with a slight relief. "Then you do have a country house, but somewhere else, perhaps?"

"We have a house . . . that is, we rent . . . well, our other house is in town."

"Ah yes, yes. I see." The matter of our solvency having been settled: "I hope you will forgive my silly questions, but you see, I am so . . . so new to all this, and it is so different. For example, your grass. It is so lovely and tall. I don't know whether you cut your grass here or not, but we do cut ours in England, even at our country houses."

"We cut our grass too," I hastened. "Our neighbors, the Johnsons, their grass is cut."

"But of course. Yes. I see. But yours seems so . . . er . . . so fine and tall, I merely wondered if . . . if you preferred it that way . . . whether it was a custom, you know."

"No," Elisabeth put in firmly, "the custom is to cut the grass, but I rather like it tall."

"Oh, but my dear, so do I, so do I!" her old friend agreed, "and it's so . . . so *different*. And do you have it tall at your . . . your other house?"

"No," I said. "We are forever cutting it there."

"Ah, but you like it being tall here, do you?"

"The truth is that this is rough ground for a mower," I confessed, "and what with the rocks and hummocks. . . ."

"I expect it would be quite difficult," she concurred, "quite." She gazed about to examine other things. "Remarkably fine view from here . . . just remarkable! Pity that great huge tree is so in the way there."

"That's our willow," defended Elisabeth. "I think it's beautiful. Have you ever seen such a big one?"

"Gracious no, never such a big one as that! But then, everything is big over here, isn't it? Bigger and better. Isn't that what you folks always say, Bob?"

"No. Not really . . . not always," I squirmed.

"Get down off the wall please, Mark," said Elisabeth with restraint.

"But Richie dropped my car down into the bushes!" Mark complained.

"He *gave* it to me," said Richie hotly.

"Did not!" cried Mark, his shrill voice rising another couple of notches. "He took it, Mom; he took it and dropped it, and on purpose, too!"

"How about a swim, you two?" Elisabeth sweetly suggested.

"He's gotta get my car. Dad, make him get it!"

"I'll get it, Markie. Never mind. But you two have a nice swim before dinner."

"I want my *car!*" Mark shouted.

"Come on, kids," I said, getting up, "let's go. Nothing like a swim," and I took each firmly by the hand and marched them down to the river.

"Splendid little fellows!" Mrs. Blanchley-Hoyt congratulated as her voice faded behind us. "I remember when you and Brenda had such times together. . . ."

"You can't just *leave* like that!" Elisabeth whispered angrily in the kitchen as I collected plates to set the table. "She's a guest. You've got to entertain her."

"Just getting the splendid little fellows out of the way," I explained. She growled.

"Not with your knife, Richard. Use your fork," Elisabeth encouraged at dinner. Then, to Mrs. B.-H., "Table manners are different here."

"Yes," agreed Mrs. Blanchley-Hoyt.

"Just conventions—that's all they are," Elisabeth went on. "Different in different places. Just a matter of what we're used to."

"Quite. Quite, my dear," said our guest somberly. "I do think sometimes that we are actually *too* civilized in England. We lack your . . . your American dash, Bob." I shifted uneasily. "We just don't have that . . . that *je ne sais quoi*. Pity. But we are so isolated, you know. I've always felt that . . . always. I read just packets of things . . . absolute packets before coming, but these people . . . why, they spend two weeks in another country and then write books about it! Do you still write, dear?" Elisabeth muttered something. ". . . because I can remember how you and Brenda were always *so* literary . . . always scribbling! Well, not scribbling, really. Just a wee, tiny slip of a thing she was, perhaps she was thirteen, but so proud, terribly *proud* when she announced that she was writing a book. And I don't expect you would remember, dear, but she actually did. Finished it too. She writes, you know, for *Twentieth Century* now. Can't think where she finds the time, what with all those dear little ones, but she does, and her editors say she has such a clear style. Bucks her up, quite bucks her up to hear things like that, and they are thinking of giving her a regular column. What sort of things do you write now, dear?"

"Oh, just a little of this and that," said Elisabeth casually. "Knife and fork on the side of the plate, Mark. No. Like *this!*"

"Brenda and Philip . . . oh, they will be pleased, so pleased to hear all about you and this . . . quaint place!"

"I'm awfully sorry," broke in Elisabeth, "but I must go to Kingston tomorrow. I have a doctor's appointment, and I couldn't change it."

"Oh, of course you couldn't. Don't mention it. Of course you must go. Will Bob be going with you?"

"No. I'm going to be here. As a matter of fact, I've got to fix the roof. A couple of shingles blew off, and Jack Stuart is coming over to give me a hand."

"Ah, I see . . . a roofing person, is he?"

"No. Actually he's a prison guard, but he's a friend. He knows how to shingle and I don't."

"How very interesting!" she exclaimed. "A guard at a prison, and he can do roofs!"

"Yes," I answered.

"The children will come along with me," Elisabeth said, "so you two will have a nice quiet day."

"We shall manage famously, shan't we, Bob?"

"Indeed we will," I said.

A few minutes after enviously cheering Elisabeth off next morning, I went upstairs, climbed out the bathroom window onto the porch roof, and hunted around for the leak. Of course I couldn't find it, but it was lovely up there with Mrs. B.-H. hunting out fresh weak spots to focus on below. After a few minutes she suspected my malingering and called up, "Getting on up there, are you?"

"Yes, yes I am," I hastened, rising to a half-sitting position and thumping the roof casually. As I relaxed again, I cautioned myself to thump every now and then. "A little hard to figure out though," I confessed after a few minutes and a few more aimless bumpings. Off in the distance, toward town, I caught the hum of an approaching outboard. As its speed slackened and it turned in toward the dock, I got up, climbed back through the window, and went down to meet Jack.

"How do you do, ma'am," said Jack, instantly a bit sheepish.

"Do sit down, Mr. Stuart?" Reluctantly, he did. "Bob tells me you're a security officer at a penitentiary."

"I guess you could say that, ma'am," he replied.

"I find that so interesting," she said with enthusiasm. "For a few years I did—was involved with, I suppose you'd call it—social work; so I do find that sort of thing most interesting indeed."

"I see," he answered.

"The sort of people . . . I am interested in the sort of people you work with. How would you describe them?"

"Describe them?" Jack's voice was puzzled.

"Yes. How would you describe them?"

He considered for a moment.

"Well, ma'am," he finally said with a nervous laugh, "the long and short of it is, they're criminals!"

"Yes, yes of course, . . . but the sort of person? I mean, are they . . . are they what you would call well-to-do . . . or . . . how shall I say . . . perhaps working class?"

"We got all sorts," he said cheerfully; "any sort you'd like, really."

"But your system," she went on, switching to another tack, "your prison system is very different from ours, I dare say. What is your system like, exactly?"

He wrestled with the problem for a while. "Well, have you got walls around yours?"

"Oh, quite! Yes, walls."

"Sounds like the same system, then," he declared with satisfaction. Annoying as I found Mrs. B.-H., I felt a few pangs of sympathy for her as she tried to regroup her thoughts under Jack's unintentional rebuffs. Then, too, I felt sorry for him.

"Excuse me," I finally interrupted, "it is bad form, I know, but shinglers are hard to find. . . ."

"Free ones are," Jack said, rising with a laugh of relief. "You got the shingles and the gunk?"

"If you'll excuse us," I said, "we'll be just overhead."

"Not such a bad spot if it comes to that," Jack observed, relaxing on the patched roof with a cold beer. We sipped and considered things for a while until a thought struck me.

"You wouldn't like to run me into town, would you? I need a part for the toilet."

"Sure," he said; "let's go." After a rather embarrassed explanation of the mission to Mrs. B.-H., closing with a couple of repetitions of how I would be back soon, we took ship and pushed off. It took only a couple of minutes to pick up the part at the hardware store, but we had to stop in at the International for a "draught," as Jack put it, on the way back. Gently and sweetly the draught and the river talk flowed as the hands of my watch turned themselves slowly through the afternoon.

"Jack," I said irritably, "for years now you've been promising to pick me up a young crow or a raccoon. If you don't start producing pretty quick, I'm going to start losing faith in you."

"Not yet I haven't, and, what's more, you ought to be damned grateful. Too late now this year anyway. They're all finished nesting, and if you don't get 'em right when they're little wee fellas, you'll never tame 'em. What the hell you want of a raccoon anyway? He'd rip your place to bits. Climb right up the walls, they do, and whip the shades and curtains right off the windows. Terrible little buggers, they are. Hey, Phil," he called to the waiter (in Canadian pubs it is illegal for a customer to go to the bar for his beer. He is required to sit and be waited on at a table). "Give us two more here."

"They couldn't be that bad," I countered; "anyway, I'd just like one, that's all. Anything wrong with that? I like animals." Phil plunked the beers in front of us.

"Bob here—he's nuts, you know—he says he wants a young coon. Know anyone who's got one, Phil?"

"Want a coon?" Phil asked as he was picking up the money.

"Yeah," I answered. "Know where there is one?"

"I got one you can have," he said. My heart gave a sudden jump.

"You really have a raccoon?"

"Sure," he replied.

"What do you want for him?" I asked urgently.

"Nothing," he said.

"Give you five bucks."

"Nope. When do you want him?"

"I could come in tomorrow, I guess, but I'll give you . . ."

"If you want him, you can have him right now. Got him just up the street here. Hang on and I'll bring him down."

As if stunned, I sat trying to absorb the idea that soon I would be holding a live raccoon—not just listening to a description of one from a sighted friend, not running my fingers over a rigid, immobile statue of one covered with plush. Soon I would be in touch with the living fact. A visit to Krueger would be for me only a long, dreary ride over bumpy roads, an experience not to be compared with the touch of a raccoon's paw, the sense of those instant, razor claws, lightning-swift, dexterous, wielded by round, strong, but not heavy shoulders under that inch of warm fur. To touch these, to sense the life animating that superbly competent flesh, would be to experience the impulse which sends that creature drifting with air-borne stealth through the tall grass along the river in the late evening, sniffing, listening, looking, waiting, watching, sifting through shadows, alert, tense, alive. The touch, the sound, the smell of that animal would be my trophy room, my zoo, my instant pathway back to my feral origins where the pulses still throbbed in the dark and whispering branches.

This and all that more which must ever escape mere statement set my heart to pounding in the beer-soaked atmosphere of the International when Phil had simply said, "Hang on and I'll bring him down."

And, I thought, with this new-old part of me giving off the rich redolence of Canadian draught, I would return to the island in triumph and introduce Mrs. Blanchley-Hoyt, saying, "This is my *real* country house," to which she would say, "How interesting! How very interesting," and she would not understand. Richie and Mark would understand. They would be ecstatic. Elisabeth, though she would not let more than the top of her pleasure show above her anger, would be happy too.

"Jesus!" Jack laughed happily, "won't you just be in for it now! Won't you just! Here you'll come rolling in full of booze and with that coon! She'll be overjoyed, she will!" He laughed again. "You know, though, I used to see one on the way to work. Fella had him staked out in his front yard, and he had this big collie dog. How them two would play together! I'd pull right off the road and sit and watch 'em. Wrastle and roll around, they would, and snarl and yip and carry on something wonderful. All in fun, you know. When that little bugger had enough, he give that collie a swipe across the nose and scoot up this little tree like greased lightning." He laughed. "Jesus, it was fun to watch!" Under the part of him that had asked why I wanted one, Jack knew, and that's why we were friends. "Used to be this guy out at the pen, you know. He had this weasel."

"What? In his cell?"

"Yeah, sure in his cell. Where the hell do you think? Found him when he was a baby no bigger'n your finger, and he nursed the little fella with a piece of string he'd dip in a saucer of milk. He'd dip in the string, you see, and the little chap would suck the milk out of the string. Christ, must have taken him hours to feed the little bugger. But of course, I guess you could say he had the time. Anyway, he did, and it grew up."

"Isn't it against the rules to have pets in prison?"

"Sure it is."

"Then how come this guy had a weasel?" Jack leaned back to laugh again.

"Ever try to catch a weasel? Like a shadow, he was. Couldn't hardly tell when you were looking right at him, and anyway, who the hell's gonna grab a weasel, I'd like to know. Them things could snip off three of your fingers while you're tryin' to decide which way to move. Nobody'd lay a hand on it even if they could've. But this guy . . . Christ, he could do anything with him. Used to carry him around inside his shirt, but couldn't nobody else touch him."

My heart went out to that prisoner, whoever he was, and I was glad for him. Though mine were different walls, different bars, we were brothers. Jack understood too.

"What ever happened to him?" I asked.

"Oh," he said with a trace of wistfulness, "who knows, eh? Who knows?" After a pause he roused himself, "Oh Jesus, here he comes!"

"Has he got the raccoon?"

"He has that," Jack assured.

Phil came through the sudden commotion and said, "Just put your arm out here, Bob, and he'll step on. Don't try to handle him though. He's nervous."

"What's the coon got to be nervous about?" laughed Jack. "It's Bob's gonna lose an ear or so!"

Claws pricked through my shirt cuff and, softly snarling, on tense legs, the raccoon walked up my arm to stand on my shoulder. My right ear tingling, I slowly moved my right hand into my pocket and pulled out a bill. "Here, Phil."

"Buy me a beer sometime," he answered and was gone. Chairs scraped back and feet shuffled as people moved in for a closer look.

"Yours, Jack?" someone asked.

"Christ, no," Jack rejoiced. "This here is Russell's problem." A cold nose sniffed at the tingling ear. "I'm not moving," I argued silently. "I'm supposed to keep still, and I'm keeping still, but do you know the rules?" Though my fingers itched to examine him, I kept my hands firmly on the table. A group began gathering about my chair, and, as it did, a soft gurgling in the coon's throat grew louder; I didn't move even harder. On tiptoe—all claws—he picked his way around the back of my neck and onto the shoulder nearer the wall, and my left ear began tingling.

"He's tame, is he?" someone inquired.

"Christ knows!" Jack chuckled.

"Hope so," I managed. They moved closer. Through the needles on which the racoon stood and moved, I could sense his tension mounting. "Jack," I said quietly, "it might not be such a bad idea if we got . . . if we moved very slowly out of here. I think all this noise bothers him," and I began gently rising. It was too late.

He sprang from my shoulder to the table, onto the floor, and was gone through a shout and a scuffle.

"There he goes!" shouted Jack, leaping after him, and there was a rush as others joined in the chase. I swore at my inability to help. "He's going to get away," I muttered. "He's going to get away!"

"Take it easy, now," Jack's voice commanded from another room.

Skippie, the owner, called, "Shut that pantry door! If he gets in there with all that food . . ."

"Whoop! There he goes!" everyone chorused, and the thunder of running feet swept into the ladies-and-escort room where a woman screamed.

"Keep him out from under!" someone shouted. "Oops! Too late! He's gone." I crumpled up with misery. Gone, and before I had even touched him.

"Can't reach him," Jack cried. "He's way back in the corner there. Boobie, come here. Your arms are longer'n mine. See if you can reach him."

"Go to hell," growled Boobie Woods. "Catch me reaching under there for a cornered coon!"

"All right, then. Skippie!" Jack called, "give us a hammer and a wrecking bar. I'll have to pry up the seat of this bench."

"God sake!" Skip objected, "I just put that in last week, and you aren't going to do any hammering and prying on it!"

"Serves you right, making stupid improvements around the place. How the hell is a fella going to get his raccoon out from under a bench all closed in like that? Trouble with you, Skip, is you just don't think far enough ahead, that's all. You just run along and get me that hammer and bar, or, by God, I'll leave him right there, and how'd you like that, eh?"

Skip groaned, and soon there was the sound of hammering. "Easy does it, Jack," Skip pleaded. The hammering stopped.

"Now, just you don't interfere between a fella and his work," Jack threatened, "or, like I said, I'll leave him there."

"But Jesus," Skip worried, "just take it easy, that's all."

"I'm doing my job," Jack went on. "You stand back there and get ready to do yours. When he comes out of here, teeth bared and claws swinging, you're supposed to grab. Understand?" He returned to his hammering. When he got a bite with his bar under the seat, nails screeched as he pried. A moment of silence, then a shout of victory.

"Now you get him the hell out of here," called Skip. Jack came back to the men's bar with the raccoon on his shoulder.

"Come on, Bob. You know, I don't believe we're so popular as we were a little while ago. Hey, Skip, bring us a box for the little fella, will you? I'm not just gonna throw him into my car like this!"

"Throw him the hell in and take him wherever he's going," Skip urged.

"Can't do that!" Jack objected, "we need a box. Besides, I'm not done drinking yet." Skippie disappeared and returned with a big grocery box, and Jack popped the raccoon inside. I carried it to Jack's car and put it in the back seat with a rock on the top.

"You take it easy in there for a bit," Jack advised the raccoon, "while we have a few beers. We'll be back." We were seated once more at the table when a woman approached us.

"Whose coon is that?" she inquired.

"Why, that's Bob's coon," Jack said.

"I'll give you fifty bucks for him," she said.

"He's not for sale," I said.

"Everything's for sale," she declared. "Make it a hundred. I had one when I was a kid and I'd give my left arm for another."

"Sorry," I said.

"Last offer, then—a hundred twenty-five."

"Sorry. He really isn't for sale," I said, and she went off swearing to herself.

"There's somebody's crazy as you are," Jack said as he ordered two draughts. "On second thought, maybe she isn't. You could've just made yourself a nice piece of change. Wha's the matter with you?"

"Dumb, I guess. But what's that?" There was a sound of more hammering from the other room.

"Just the boys fixing up that bench," he said, pouring salt into his draught; "They're a good bunch, but they make too much noise to be good coon hunters."

By the time we got back to the island, dinner was on the table, but it got cold. Elisabeth postponed her explosion, but the kids had theirs and enjoyed it.

"Is he real?"

"Can I pet him?"

"What's his name? Where did you get him?"

"Leave him alone, Markie. It's my turn."

"Stop! You're making him nervous. *Quit it!*"

"Move over!"

"Get out!"

"Dad, make him get *out!*"

"Let *go!*"

"What was the name of that animal?" inquired Mrs. B.-H. "Did you say it was a coon?"

"Raccoon," I answered carefully, "very fine animal. He's my guest from a far countrie, or maybe we're his guests. Anyway, he's long overdue, and I've missed him."

Exciting as was his welcome, the raccoon seemed to find it less disconcerting than the cafuffle at the International. Though preserving the dignity of his person, he consented to visit the shoulder of each—except that of Mrs. Blanchley-Hoyt, who extended no invitation.

That night I attached him to a spindle of the porch railing by a long dog chain and collar, gave him a grocery box padded with rags, put down a bowl of bread and milk, and went to bed. I began drowsing off with Elisabeth angrily whispering into my left ear the duties of a host to his guest, about how shameful it had been to leave Mrs. Blanchley-Hoyt and go off boozing with Jack, and elegant variations on similar themes. As she hissed along, my right ear tingled with the touch of the raccoon's whiskers, and I heard the strange cluckings and gurglings from his throat. Though drifting toward sleep, I caught myself and went down to make sure he was all right. He greeted me with a soft snarl from his box and nipped my fingers in warning.

"That's all right," I said. "I don't blame you for being nervous, but you'll find that we're friends and I'm glad to have you with us." And I returned to bed and slept.

Next morning, early, I went down to see him. His box was empty. I found the chain. It was stretched taut along the edge of the porch. I followed it along, calling softly to him. The chain wound in and out and around the spindles, and I traced it carefully until I found him at the end. At my touch, he neither moved nor made a sound. Head wedged between two spindles and secured there by the twisted chain, his body hung limp and lifeless over the edge of the porch. In that morning full of sun and birdsong I held the dangling body in my hands. Only yesterday he had come to me out of the forest of my ancient past and, after one evening, had left the inhospitable countryside of my life.

"It's not all right," I said, undoing the hangman's noose of a collar; "not all right at all. I didn't mean this—not that that matters now—but I didn't." Lovingly I traced the joints of the shoulders, examined the still paws with their razor cleats, opened the mouth, felt the teeth, studied the skull, ears, and down the ridge of the backbone to the drooping tail that would arch again only in my imagination. I strove with reverence to learn all I could learn, but all I could then learn were the useless things. The raccoon had gone. All that remained was a stuffed museum piece.

Laying it gently in the box, I went back to the kitchen where Elisabeth was cooking breakfast.

"He's dead," I said. Her tears were sudden and soon gone. She made pancakes to cheer the heartsick children.

"I did hear this very odd sort of yowling in the night," commented Mrs. Blanchley-Hoyt, "and I did wonder what it could be."

"I wish you had waked us," I said.

"Yes. Quite. But then, one doesn't know, does one?"

"No," I agreed; "one doesn't."

"But, my dear, Brenda and Philip will be so interested, so interested when I tell them about this, and these . . . these . . . what do you call them again?"

"Pancakes?" suggested Elisabeth. "I'll get some more," and she went to the kitchen.

"Ah, Bob, they are delicious! I expect she made these for my last breakfast to prove she could actually cook, don't you think?"

"Exactly so," I said.

And she did leave that day, with her freight of interesting news for Brenda and Philip about our country house with its uncut grass and poor dear Elisabeth's cooking. Her going was a shaft of light in a day darkened by the departure of our other guest.

Toward the close of the season we were sometimes disturbed at night by the sound of an animal running about in the attic. On the last night of that summer, having brought in the fiberglass, hung the motor on its sawhorse, and put up all the shutters, Elisabeth and I relaxed on the front steps in the light of a full autumn moon. Behind us the house stood silent, cluttered, clenched against the coming winter. Elisabeth took a few steps toward the dock and turned to look back at all we were about to leave, the cold and silence already beginning to creep into the house through the chinks.

"Bob!" she cried with excitement, "Bob! It's a raccoon! It's a raccoon up on our roof. He's just sitting there in a little hole over the attic window, and he is positively *waving* at me! You've got to close off that hole. It's him we've been hearing, then. We can't just leave it like that. You've got to do something!"

"He won't hurt anything up there," I said a little wearily. "Besides, he needs a place for the winter."

Getting up, I walked toward the river, then turned and called up, "It's all yours! Be our guest!"

8

Air Guns

"But that's all I want is an air gun," said Richard in Lancaster one fall. "I'm eleven, and you said I could have one when I was eleven."

"I said I'd think about it," I corrected.

"Well . . . have you?"

It was my turn to be uncomfortable.

"I've thought about it, Rich, and there's one thing definite. You can't have one for Christmas. It's just no good having an air gun in town. I know." Memories of my own air gun were still fresh after twenty-five years. The tingle in my arm as I reached under the pile of old pillows in the attic where I used to hide it; then my fingers touching the smooth wood of the stock, tracing my carved initials, then up to the cold, quiet steel of the barrel. I remembered, all right. But I also remembered my secret atrocities —the spider-web cracks in neighborhood windows, the squeal and slap of a woman up the street as she grabbed at the imagined bee. My parents never knew, never even suspected. But now I was the parent. Rich had asked openly and honestly, showing more courage than I. My parents would certainly have said no. After all, who puts an air gun in the hands of a twelve-year-old blind boy? So I had to get it my own way.

"They're just no good in town, Rich. If you had one, you'd want to use it, and you just couldn't. Besides, there are laws."

"I guess so," he admitted sadly.

"I know how you feel. It's hard, but it's only sense, isn't it?" Oh! the weight of my common sense!

"Still," I said, "look. We'll be on the island next summer and maybe . . ."

"Yeah?" his voice rose, "that's not town. There's the whole St. Lawrence around us. Yeah. What about it? Can I have a gun up there?"

"Well," I tapped my cigar thoughtfully on the ashtray. He waited. "I guess you can. That's a good place for an air gun, but you'll have to buy it yourself."

"Yippee!" He began leaping around the kitchen.

"But listen, Rich," I said in the soberest voice I could muster . . .

"Yeah? Yeah?" He leaned tensely across the table.

"A gun is no plaything. There's Mark, James, and Miranda to think about with a gun around the place, so . . ." and I launched into a ponderous lecture on the danger of firearms, with Rich trying not to bounce.

All winter and spring the pile of small change slowly grew in his bottom drawer. About the middle of March catalogues from mail-order houses began sifting through the letter slot, and I would come home from the college to find the family hunched over a new one with a smell of burning from the oven.

Rich met me at the front door late one afternoon in mid-April.

"Here you are," he said, "hold out your hands." And into them he poured about a pound and a half of nickels, dimes, and quarters. "Six-fifty. Make out the check. I've got the order slip."

"But wait a minute," I said, putting the treasure onto a side table and taking off my coat. "How do you know that's enough?"

"Six twenty-three it costs. Twenty-seven cents for postage. Says so in the catalogue. Want me to read it to you?"

"No, but what kind are you getting?"

"Winchester repeater special."

"But the army deluxe! that sounded like a better gun. It had . . ."

"Nope. Winchester repeater special is the one I want, and, after all, who's buying it?"

Two weeks later the mailman brought a long, thin parcel.

Through the thick brown paper and heavy cardboard I could smell, could taste air gun. I resisted opening it, though, until its rightful owner returned from school.

Mark skipped back and forth through the front door while we waited for Rich. When he caught sight of him he shouted, "Hey Rich! Your gun's here! Came this morning."

Richie began to bound up the front steps, then caught himself before he had entirely shed the dignity he felt properly belonged to an older brother. Walking slowly across the porch, he stamped his feet a few times—a pretense of shaking off some nonexistent mud—came in, coughed twice, asked huskily, "Where is it?"

"Dad's got it in the living room."

"He didn't open it!"

"Course not," I defended.

He took the package from me tenderly and, breathing quickly, began to strip off the paper.

"Oh!" cried Mark in disappointment as he hopped up and down. "There's a box!"

"Course there's a box, you silly idiot," said his brother. "There's always a box. Do you think they'd send it without a box? Get damaged or something."

"Well, they might. Hurry up anyway, will you?" Richard slowed down accordingly and began painstakingly to strip off with his fingernails the heavy gummed paper that sealed one end. Digging in my pocket, I pulled out my penknife. "Here, try this."

"That's all right," he said, "I'm getting it."

"Yes, but this'll be quicker—easier."

"No. Wait a minute. I've got it," and he freed the narrow lid with its long tongue. I was on my feet.

"Let me hold the box for you while you get the gun." Mark and I held the box while Richie reached in.

"Got it?" asked Mark.

"Not yet, but there's something . . . wait! Here it is!" The box came to life in our hands, quivering as he withdrew the gun. When he saw it, even Mark stopped jumping. He was frozen with awe. As Richie's fingers caressed the stock, the cardboard in my hands seemed to contract and harden into the smooth, old, familiar stock of those other years, and my flesh on his bones trembled again.

Richie was the first to jump from the water taxi that June onto our dock, carrying his long box, unsealed at one end. At breakfast on our third morning, after two days of incredible patience while we settled in, Richie said, "Well?"

"Okay. After breakfast," I agreed. "We'll go out back . . ."

"But Richard," interrupted Elisabeth severely—

"Yes, Mum?"

"No birds! Remember. I won't have you shooting things with that gun."

She seldom left the house without her binoculars for fear of missing a new bird or even a good look at one of the redwings, song sparrows or yellow warblers that were so plentiful. I was as interested as she and went along on the bird walks as book-bearer, locating them by sound and learning their calls. Though Elisabeth was enthusiastic and observant, she was no match for Richie. He was annoyingly quick at spotting and maddeningly exact in identification. He was our naturalist, in fact, with a collection of paperbacks on natural history that would do credit to an amateur twice his age. Considering all this, Elisabeth's warning seemed unnecessary, but she continued, "I don't want the birds frightened away from around here. Do you understand?"

"Sure," said Rich. "I understand."

At six he had snatched his first bird from the jaws of Sinbad, our fierce male Siamese, a feat of agility and nerve that can only be appreciated by one who has tried it. Young rabbits, moles, mice, and snakes followed, as if he had some hunting agreement with Sinbad. So, since he had struggled to revive and keep what Sinbad had caught, Elisabeth's fears seemed groundless.

Forty feet from the back porch and parallel to the house ran an old wire fence, and beyond stretched a third of a mile of tall brush and undergrowth separating us from the empty eastern shore of the island.

We all trooped onto the back porch, Rich carrying his gun. I carried the beebees. I got some putty from the shed, leaned an old shutter against the back fence, and slapped the putty on the shutter for a target. This done, I herded the other children onto the porch and threatened immediate and terrible reprisals if they stuck so much as a toe over the edge. Then I turned and loaded the gun, preaching all the while. Like most teachers, I was well up on principles but weak on practice. My experience was no use at all. It consisted of listening carefully and then shooting at whatever made the noise. I understood—and so did he—that the point on the peak of the barrel should be lined up with the slot in the rear sight, and I spun this out in my best platform manner. He bore with me patiently. Then, finally, "Okay?"

"I think that's about it," I said.

He took the gun and, to show the spectators that I obeyed my own rules, I stepped behind him.

Puff—tock!

"Bull's-eye!" shouted the others from the porch.

"Hurrah for Richard!" added Mark with a howl of admiration.

My rules about the storage and use of the gun were so stringent and effective that the novelty of its presence gradually wore off for everybody except Richie—and me, of course. The placid unroutine of our days continued—swimming, fishing, trips to town for groceries, Elisabeth reading aloud to the family on the front porch on drowsy afternoons with the teapot on the painted table at her elbow and everyone silent, crowded around, forgetting the sunshine and the river and his tea—everyone, that is, but Rich. He was alone, off in the brush somewhere with his gun, doing and being me. And the cries and rustlings that crackled in the boy's ears, stirring the winds that sang through his blood, crackled and sang through his father.

Soon no one bothered to ask, "Where's Rich?"

Late one afternoon I was cleaning and oiling a reel at a table in the living room. Through the open window beside me a light breeze carried the voices of Mark, James, and Miranda as they splashed in our shallow bay. Elisabeth was getting dinner ready, and above the kitchen noises I could hear the "pft!" of Richie's gun out back. Elisabeth stopped what she was doing and came in. "Do you have to do that there?" she asked.

"No," I admitted. "It is pretty messy, I guess."

"I don't mean that," she said. "It's just that Rich is out there with his gun and, well, I get nervous. Couldn't you do that on the back porch?"

"Sure. Good idea. You're quite right. He's only eleven, and I did say . . ." I picked up the reel and bottle of oil and went through the kitchen out onto the back porch. Elisabeth returned to the dinner.

"Hi, Rich?" I called.

"Yes, Dad?"

"What are you shooting at?"

"Oh . . . post."

"Let's hear how good a shot you are."

He cocked the gun, scarcely seemed to take time to aim, and fired.

Tock!—the beebee cracked into the post forty feet away.

"Good shot!"

"That's an easy target," he said.

"Try some harder ones, but be careful," and I turned my attention back to the reel. He went on firing.

Suddenly he shouted, "Hey," and his voice dropped, surprise filling it. "I got him."

"Got what?" I asked absently.

"The sparrow. There was a song sparrow on top of the clothes pole. And . . . well . . . it was a hard shot—a real hard shot, so I tried. And he just fell down. They just fly away, you know; even when I hit them they fly away, but he didn't." Awe, surprise, dismay crowded around the edges of his words.

"Well," I said, putting down the reel and stepping off the porch, "you probably just winged him, you know—stunned him."

"Maybe," he said doubtfully.

"There's not enough power . . . go and have a look and see if you can find him."

"Here." He handed me the gun and walked slowly across the yard into the tall grass by the clothes pole. For a long while he stood there, parting the grass and looking. Then, "Here he is," and another long pause.

"Bring him here. Let's have a look."

Slowly he wandered back toward me. Stopping halfway, he said, "I think he's dead, Dad."

"No. He's not dead. Let's see." And he came to me.

"I think he is," he said quietly.

I put the gun down and took the bird. It was warm but still, and the head sagged over my forefinger. I put the tips of my fingers against the breast. Not a quiver of a heart. I pulled gently at a wing. Limply, it accordioned out.

"I never realized they had such big wings, Rich."

"Yes," he said abstractedly, "they're big, all right. Do you think he's alive?"

"No. Not really." We were silent.

This was perhaps his first moment of conscious paradox—his pride in the shot and his misery at its consequences, his love of the sport and his love for the bird. He faced the contradictions of his human nature. How should, how would he react? Cry or pick up the gun? Perhaps he was afraid, too, of what I might say to him when he was so vulnerable.

And what should I say? I wanted to say the right thing, the thing that would help him, guide him. What was a father for if he couldn't do that? He was waiting.

"You think it's a song sparrow?" I had to say something.

"Yes," he said, "the markings on the breast. It's a song sparrow." Beneath the words was the sound of tears.

What I should have said—what was right to say—I didn't know, never would know, but I did know there was a part of him calling for help.

"Well, Rich, Audubon, you know, was an excellent shot, and he used to load his gun with just enough powder so that he stunned the bird. He had to, or else he never could have seen them as close as he needed to. And sometimes . . . naturally, sometimes he killed them by mistake."

"I know," he answered.

"Where did you hit him? He doesn't seem to be bleeding."

"Must have been under the wing. There's a speck of blood, but when I hit the others they flew away."

"I bet this one must have spread his wings just at the wrong time. Feel those wing feathers. When they're folded, those feathers make a really tough shield. A beebee wouldn't hurt them. It would scare them, but not really hurt." He tapped the wing I had folded back into place.

"See? That's pretty tough—even tougher on a redwing or a flicker. I bet you've never seen a song sparrow so close."

"No," he said, taking him tenderly from me. He examined him again in silence. "He just must have spread his wings, that's all."

"That's what happened," I said. "Do you want to keep him?"

"Guess not. What should I do with him?"

"Throw him away. A crow or a gull will pick him up and be grateful."

He walked to the fence and threw the sparrow far into the bushes, paused for a moment, and returned.

"Can I have the gun, Dad?"

"Sure. Be careful, that's all." As he picked it up, I went back into the house and sat down at the kitchen table. "How's dinner coming?"

"Fine," said Elisabeth from the stove. "Be about ten minutes."

"Richie just shot a sparrow."

"He didn't!" she exploded. "Was it hurt?"

"Killed it," I said.

"Did you tell him off?" she asked angrily.

"No."

"Why not?"

"I don't think he needed it, and anyway . . . I couldn't."

She walked to the door and looked out. "He's still shooting," she said.

"I know. Don't say anything."

She opened the door, walked out onto the porch, paused, and came back.

"Here's your reel," she said, putting it down on the table. "You must have forgotten it."

"Thanks," I said. "Thanks."

9

Captain Outrageous

I lay on the dock drinking in the peace of a quiet Saturday afternoon. Though it was mid-June and the sun was warm, the shadows kept alive the memories of recent frost. Until July we would need two blankets at night, no matter how tropical the afternoons.

Miranda and James, in bathing suits and life jackets, fished beside me. Miranda more than kept up her end of a friendly one-way conversation with the rock bass and perch nibbling at her worm. James sat silently dreaming the dreams all fishermen have dreamed since time began, and I basked in the sun smelling the perfume of the cedar planks beneath me, listening to the drone of an occasional outboard and the cries of a solitary gull.

"I've got one!" cried Miranda, jumping up, and a rock bass flipped and flopped on the boards beside my head. Without getting up, I reached over and unhooked him.

"Let *me,* Daddy; let me throw him back." I held out the fish. "No. Put him down and I'll pick him up," she said, backing away a little nervously. I dropped him and squatted to give him instructions about where the edge of the dock was and to help him along in the right direction with a tentative finger. Plop! and he was gone. For a long moment Miranda considered things. Then, "When will Mom be back?" Elisabeth had left around noon to drive the hundred twenty-five miles to Syracuse to pick up a guest, and she had taken Richard and Mark for company.

"Late," I answered. "You'll be asleep." The wake of a passing boat slapped against the side of the dock.

"But won't it be kind of hard coming back at night?" Miranda pursued. However well one knew the river, crossing from the mainland in the dark was a tricky business. Everything looks different after the sun goes down. More than one of my friends with a lifetime of experience on the river had run their boats onto rocky shores that they would have sworn in daylight would have been open water. So frequent are these mishaps that some natives swear that, come nighttime, the shoals and islands move about like uneasy whales.

"I'll put the dock light on," I said. "She'll be all right. Getting hungry yet?" My watch said five o'clock.

"A little," answered Miranda.

"Me too," said James.

"Okay. Come on, then. Let's go up to the house and see what we can find. Reel in, and—"

"But Dad," James interrupted, "there's a boat coming in here."

"Can't be," I objected. "Who'd be coming in here?" But there was a boat, a large inboard, by the sound of it, angling in from the main channel. A mournful mechanical "moo-oo!" from its horn announced her intention of docking. It sounded like the water taxi, I thought with dismay. I struggled with my memory to recall all those rash invitations issued thoughtlessly toward the end of cocktail parties last winter. It could be almost anyone! But "anyone" would almost certainly have a large family—he always does—and he would leap ashore with a hearty handshake and begin introducing innumerable small children. I did my best to look pleased, but it had been such a pleasant afternoon!

The powerful engine came on with its gentle, deep-throated muttering as the big craft slid across the face of the dock.

"Hey!" James cried, "it's Tom!" Tom? Tom? Tom who? The motor died.

"Hi, Rob. Get that stern line, will you, Bill?" I recognized the voice. It was Tom Murphy, my brother-in-law from Syracuse. An excited phone call from Hunnie in the spring brought the news that they had bought an island over on the American side with a large house, a larger boathouse, and, if memory served, a small fleet.

"Tom! Welcome!" I called with relief. Feeling my way along the dock, I said, "Give us your bowline." He tossed it up and I made it fast. "Come ashore."

"Why don't you come aboard?" he countered. "My brother Bill's here." We shook hands.

"Is this your boat?" inquired James.

"Sure is," Tom proudly declared. "Come on aboard, all of you. Where's Elisabeth? I thought we might give you a ride. Take you over and show you the island. It's only a twenty-five-minute run in the *Jackstraw* here."

"She goes like the wind," Bill added.

I climbed down into the broad deep cockpit, and James and Miranda scrambled after.

"Wow!" James confessed, unable to restrain his praise in the face of such magnificence. Wow indeed! I thought. To a family reared in small outboards, here was power and majesty. I explained Elisabeth's absence as I made my way forward beside the engine housing.

"Too bad," said Tom sadly. "Come on forward. Watch out for the canopy roof." I kicked something. "Just a paddle," he apologized.

"A paddle!" I laughed. "What are you doing with a paddle?"

"Regulations," Bill replied from the stern. "But I'd hate to have to use it."

"Great exercise paddling a twenty-six-foot inboard," said Tom. "But how about coming over to the island for dinner? We'll have you back before dark."

After a bit of polite backing and filling, I was headed for the house to get sweaters for the children. The river always turned cold in the evenings.

"And if you've got any Canadian ale . . ." Bill called after me.

In a few minutes we were roaring along, ale in hand, leaving a wake we would have sworn at if we hadn't been making it. Damn tourists! we would have said. Ought to be laws against people like that owning powerboats! Ah, but it was lovely—the steady thunder of the engine behind us and the *brrr* in the floorboards as we boiled through the small waves.

"She really gets up and goes!" Tom shouted exultantly. "Thirty years old, maybe, but Bill's got her sounding like a sewing machine." When he was small, Bill never got Erecto sets for Christmas. His parents gave him power mowers the mechanics had given up on.

Before I had finished my ale, Tom eased up on the throttle and we glided into one of the slips of his huge old boathouse, where Bill's wife greeted us.

"Some boat!" James admired as he climbed out.

"And what a *boathouse!*" Miranda added as she scrambled over the side.

Ignoring their admiration, I reminded them sternly about keeping their life jackets on. The warning delivered, received, and acknowledged, I was prepared with my admiration, and heavy demands were made upon it as Tom ran my hands along the two-by-four studs in the boathouse wall that actually were two full inches by four full inches—whereas the modern two-by-four is actually one and three quarters by three and a half or maybe five-eighths. Some of the supporting beams for the second story were actually eight-by-eights, and some even larger.

"It's old," he conceded, "and needs a lot work, but they really used good materials when they built her. Come on up to the house, though." The house had been done on the same scale, with its fifty-foot living room, fourteen bedrooms, and seven bathrooms.

"Generator's kaput," Tom said as we worked gently on our ale after the tour on the wide porch, "so we don't have water or lights. Pipes are probably burst all over the show, but then, that'll give us something to do." I agreed it would if he actually wanted to get all those bathrooms operational. It seemed to me he had the essentials for island life with the *Jackstraw,* but he was realistic, knowing that he couldn't stay out on the river forever, and, after all, there were his eight small children and my sister Hunnie to consider. He and Bill had made this trip minus children to get things ready for the first summer.

Bill's wife interrupted our reflections with her cheery call to dinner, and we trooped into the huge dining room. Miranda and James took seats on either side of me and began eagerly shoveling, still in their life jackets. Maybe they weren't in any actual danger of drowning there in the dining room, but I had not much faith in myself as a baby-sitter, and they seemed scarcely to mind the bulky jackets.

"It is getting kind of cold though," Miranda piped up, her stomach full.

"Hey, I almost forgot," I said, "how about the light? Hadn't we better be moving?"

"You're right," admitted Tom after a glance at the sky. "But I think we'll be all right with the *Jackstraw*. No time for coffee, though. Let's go, Bill."

I thanked Bill's wife for the dinner, accepted the extra sweaters she had dug out for the children, and we trooped down to the boat. There were two long seats forward under the cabin roof, and I settled James and Miranda in the second of these.

"Come on up here," said Tom. "I'll show you how this thing works." A little nervously, I climbed over the back of the front seat. He showed me the gear shift, throttle, and the wheel. "Make sure she's in neutral. There. Throttle down. Now turn the key here." My fingers trembling, I turned it. After a brief groan from the starter, the engine vroomed into life.

"Cast off, Bill," Tom shouted back.

"She's free," was the reply, and I felt the soft thud as he stepped aboard.

"Put her into reverse now," Tom said quietly.

"Just a minute," I said with surprise.

"Never mind. Come on. We haven't got time to waste," he pressed.

"Then you take the wheel," I urged.

"Come on," he ignored. I found the shift and moved it gently into reverse. The *Jackstraw* throbbed as it backed slowly out of the slip. "Straight back," he encouraged. "You're doing fine."

A small hand tugged at the collar of my shirt. "Hey, Dad," came Miranda's tense voice in my left ear, "you're not going to drive, are you?" I could make no answer because I was concentrating so hard to remember where the key was so that I could shut off the motor at the instant of impact.

"James! *James!* Dad is driving," Miranda informed.

"Yeah," he replied, his stolidity obviously shaken. After a pause he went on, "You really going to drive?"

"Hard astarboard," was Tom's quiet, confident instruction. I spun the wheel.

"He *is,* James," Miranda replied with mixed emotions.

"Straighten her out now," Tom continued.

"Okay," I began, rising to move to the side, "that was very nice. I enjoyed that . . . a great thrill, and thanks, but it's your turn now. This is an expensive boat and I have no intention . . ."

"Put her into forward," he replied calmly. I pushed the gear as far as neutral. "Now look! This is all very well, but—"

"Let Tom drive," urged James.

"Come on. You're wasting valuable time," Tom insisted. "Put her in forward." Reluctantly I pushed the stick ahead and the *Jackstraw* eased out toward the channel. "More throttle. Give her some more *gas!*" he ordered. I obeyed. "Go on! That's only half speed," he objected. "Shove it right up there." I did. "That's better. A little to port now. Straighten her out. That's more like it. We're really moving now." And we were. With a tense growl the engine was hurtling us along at better than thirty miles an hour. It felt like riding a charging lion. Hunched there over the wheel, into my hands and up through my arms, from my feet up through my legs, shuddered the suppressed fury of its assault that had to, must, end in the roar and crash of arrival. I braced myself for the end till my muscles ached.

"What are you so nervous about?" Tom laughed. "Everything's all right. We're out in the Seaway. Can't hit anything . . . not a thing! Well, there is that freighter off the port bow. Re*lax!*" A freighter on the left . . . my *God!* Me at the wheel and passing a freighter on the *left!* "Come on!" came Tom's confident, happy cry, "open her right up!" I touched the throttle but she was wide open.

"Wow, Dad," called James, "be careful, Dad!" I yanked back on the throttle and felt around for an emergency brake.

"Don't be so nervous," Tom urged, "go on, give her the gun!" I pushed the throttle up and we charged forward. Remembering how critical daylight would be for their return journey, I asked, "How's the sun?"

"Just setting," was the answer, "but I think we'll be all right. We haven't got time to fish, though. Here's where we swing out of the Seaway, so hard aport." I spun the wheel. "Good. Now straighten." I brought the wheel back.

My first shock of terror—or was it joy?—had passed. No, the joy had not gone. Its taste was on my tongue. It was only that its intensity had diminished enough so that it could be distinguished from terror. I felt unmistakable and clear top-gallant delight as I turned the *Jackstraw* out of the Seaway into the smaller channels between the clusters of green islands. My thirty-five-year love of the sounds, the tastes, the smells, and the feel of water seemed unbelievably to have blossomed into almost a sight of water. Was I not driving? I had slid into the front seats of our outboards so many thousands of times that I had forgotten my anger at having to leave the engine to another, forgotten even my resignation, remembering only the habit. Often I had had to explain the motor to a newcomer, demonstrate how to start it, how to steer, how to shift gears, but, the explanations finished, I had to move forward to "balance the boat." I wasn't balancing the *Jackstraw*. I was her captain, and we were screaming across the great expanse of my beloved river. My chain mail of habit, fashioned by years of incapacity, had been ripped off by our spray and lay streaming behind us on the wind.

But it was not gone. Suddenly the *Jackstraw* leaped; the wheel spun in my hands and we came down hard, still hurtling along with granite teeth grinding along under the floorboards. In a fraction of a second it was over, and we were back in quiet water.

"What the . . . !" Tom exploded. I yanked the throttle down and kicked myself over the back of the front seat to where Miranda and James were huddled. I fought Miranda's sweater off to check whether her life jacket was properly tied.

"How does it look back there, Bill?" Tom called. The jacket was tied, but the loops through which her legs were supposed to go hung loosely at her sides.

"Nothing to worry about, Miranda," I said, fighting with the knots. I knew I had to undo the jacket to get her legs into those loops. She stood up on the seat and declared in a loud, clear voice, "This is the end. We'll all be drowned."

"Don't be so silly!" I said, trying to sound convincing, but I had to admire her Bernhardt poise. "It's the end for us all!" she continued.

"How is it back there, Bill?" Tom repeated urgently.

"Can't tell," came the reply.

"Think we can make Clayton?" Tom called. Clayton was five miles off.

"Don't know," Bill replied slowly. "Might be worth a try."

"It's all over," Miranda declaimed.

"Now, just be quiet," I said, twisting one of her legs, then the other through the loops and retying the jacket.

"What happened?" James asked calmly.

"We hit a shoal, that's all," I said. His legs weren't through the loops of his jacket either. I swore at the knots.

"Is it going to be all right?" he queried.

"We're done for, James," Miranda piped.

"What are you doing back there, Bill?" Tom called with a trace of anger.

"Checking," he replied. "Checking the water. Better head for Clayton," he said a little briskly.

"I haven't got any power," Tom shouted back. Though the engine was roaring, the big hull lay almost motionless.

"Better head for the nearest light," Bill cried, a touch of strain in his voice. "Never mind Clayton." I wrestled James's legs through the loops of his jacket and tied it across his chest.

"We're going to sink," said Miranda firmly.

"The *nearest light,* Tom; the *nearest!"* Bill cried. Tom swore softly.

"Power! I haven't got any power. Prop must be gone. Have you got water back there?" He rose and started climbing over the seat.

"Don't come back!" Bill warned. "Too much weight." If she was going down, we were all going to have to get back there to avoid being trapped under the stretch of roof.

"Nothing to worry about, kids," I said hollowly. "Everything's all right."

"It's wet," Miranda announced. I could feel the water seeping into my own shoes.

"Just a little water," I said, hoisting Miranda up and carrying her back to sit on the engine housing, which was not covered by the roof. Returning for James, I carried him back and sat beside them. Bill was working hard with a sort of bicycle-pump bailer. The water was over my shoes now. Tom cut the motor and came aft into the open.

"Grab the paddle, Tom," said Bill. "Let's get back to that shoal if we can."

"No use," Miranda remarked. "We're going down." Not a trace of hysteria in her voice. She merely seemed to feel the need to announce the facts periodically. The water was over my ankles when I took the pump from Bill.

"You paddle. I'll bail," and I went furiously to work.

The infernal pump was obviously devised to give a boat owner the maximum exercise while cleaning out his dinghy after a light rain. I thrashed up and down till my arms and legs quivered with exhaustion, and a tiny stream squirted intermittently over the side. The only other sound; the swish of the paddle.

Gently the *Jackstraw* settled deeper and the water climbed. Fortunately, there was no wind. Though locked in my futile struggle with the pump, I began to feel the size of the night that was coming down upon us—how large, empty, silent, cold. If it should happen—and it could—if this was to be the end, how unreasonable, how unfair was the lack of appropriate ceremony to acknowledge our passing. The water on which we were struggling lay cold, smooth, quiet beneath us; and if it were to lie over us, it would lie equally cold, smooth, and quiet, with the vast, empty night upon it. Perhaps in a couple of days some newspapers would carry the story, but the river would remain, moving and unmoved.

The water was up to my knees. I dropped the useless pump and reached over to pat James and Miranda.

"It's all right," I gasped, and leaned quivering against the gunwale.

Splash! Someone had gone over the side.

"Throw me a line," came Bill's tense voice. Tom dropped his paddle and tossed a line.

"Got bottom?" he asked.

"Just," said Bill, and the *Jackstraw* swung slowly over as the line tightened and Bill towed her toward a rock that broke the surface. I knew enough about St. Lawrence shoals to realize there was no guarantee of a continuously gradual slope from deep to shallow water. It could be gradual on one side, then drop off thirty feet on the other.

"How's the bottom?" I called.

"Getting better," he said between clenched teeth. After a few seconds he said, "It's going to be all right." A rock grazed smoothly along the hull under our feet—the only rock I could remember being glad to have hit. Another . . . and then another.

"Great!" Tom said with relief.

"That's as far as she'll go," Bill panted. "She's aground!" He sloshed back through the shallow water to lean against the boat.

"We're sinking," cried Miranda. "We're going down."

"What do you mean?" laughed Tom. "We're not sinking. We're *sunk!*"

"Huh?" was James's query.

"That's right," I cheered. "We're on bottom—good, solid bottom."

"You mean we're not going down any farther?" James asked.

"That's it!" I said. "Not disappointed, are you?"

"Bet Miranda's disappointed," said Tom.

"What we need now," put in Bill, "is a light to signal with."

"I've got the light," Tom replied, flicking the switch of a powerful waterproof flashlight. "Who would you like me to signal?" Not a sound over the miles of darkening water.

"I thought these Thousand Islands were supposed to be the playground of the rich," said Bill angrily. "Where the hell are they? If we're stuck here all night, we'll freeze."

"There are worse places," I remarked, remembering my thoughts of ten minutes earlier. "How about an island? Can you see one?"

"There's one maybe quarter of a mile off," answered Bill. "Be one hell of a swim. The water's like ice." After a brief discussion, we agreed to try for it, leaving Tom in the boat with the light for signaling. Bill and I would try to roust up a boat and come back to pick them up. With a brave, hollow promise to the children, I stepped over the side, and Bill and I set off.

He plotted our course by the chunks of weed, figuring they would be growing in the shallowest water. Sinking almost to our knees in mud through patches of swamp, over long stretches of boulders, and across places where the bottom was as smooth and slippery as a hockey rink, we made our way. The depth varied from our knees to our necks, but at last we clawed up a small cliff to the island without ever having had to actually swim.

"I hope somebody's home," said Bill through chattering teeth. "Don't know about you, but I'm cold—least, those parts of me I can still feel seem cold." My sentiments exactly.

"Any lights?" I asked. We were on a rough path.

"No. But they've got to be here. They sure as hell aren't out on the river." We trudged on. "Wait! Here's something looks like a boathouse." Miraculously, it was unlocked and we stepped cautiously inside. "What do you know! An old motor skiff . . . and . . . my God! we're in luck! OARS! What do you get for stealing a boat?"

"Fifty years," I observed, climbing in. "Come on. You steer and I'll row." The motor and the four inches of water in the bottom made it about as easy to row as an old-fashioned back porch. My body was one large ache by the time we inched alongside the wreck.

"You got a boat, Dad!" cried James happily, and they all piled in without quite sinking us. I went mechanically back to my rowing. Tom had brought his paddle, and with his help we headed for the nearest light.

"They're not going to be all that glad to see us," Bill said as we tied up at a dock. "How about you going up and explaining, Tom? You're good at that sort of thing." Tom hesitated.

"Let's all go," he replied. "We'll make a better impression with the kids." Guiltily, we trooped up to the house and knocked on the door. Ten o'clock on a cold Saturday night in the middle of the St. Lawrence, one has a right to be suspicious about strange visitors. The man answering the door was exercising that right.

"And what do you expect me to do, exactly?" he asked. "You want me to call the Coast Guard?"

"We'd like to come in, actually," I said. "The children have had a pretty rough time, and they're cold and wet." Seeing the children, he softened and let us in.

"You know," he said, "you're pretty lucky. This is the only island for five miles with a telephone. While you use it, I'll make some coffee."

I could imagine what Elisabeth must be thinking, having returned from Syracuse to an empty house. She was probably making arrangements to have the river dragged by now; so I put a call through to her. After two rings, "Hello," she said softly and calmly.

"Hello. We're all right."

"What? What do you mean, you're all right? Where are you? I thought you were all upstairs in bed! We just got back, and I've been making the boys creep around so as not to wake you."

"We went to Tom's for dinner and had a little accident on the way home, but everybody's all right."

"But where are you?"

"I don't know where we are. Hey," I shouted to the coffee-maker, "where are we?"

"Papoose Island," he answered.

"We're on Papoose Island," down the phone again.

"Where's that? I'll come and get you."

"Go to bed. One accident's enough. We'll be home sometime tomorrow. Don't worry, that's all."

"I wasn't worried," she said.

"Well, you should have been. Good night."

"Good night," she laughed, "and no more accidents."

When Tom had discovered where we were, he called a water-taxi service in Clayton, and, after much hemming and hawing, they agreed to come out for us. While we waited, our host got blankets for the children, who fell instantly asleep. He plied Tom, Bill, and me with scalding coffee and apologies.

"Wish I could offer to take you home myself," he said, "but frankly, nothing could get me out on that river this time of night, and there's a fog coming in. But if you hit Gull Rock, how in the name of God did you get over here?" We told him the story.

"Then we borrowed a boat, an old half-sunk motor skiff," said Bill, and he tried to describe the place from which we had taken it.

"I know where you mean," answered our host, "but that boat! From the way you describe it . . . where is it now?"

"Tied up at your dock," I answered. "If you wouldn't mind, we certainly would appreciate it if you could take it back in the morning. . . ." He was on his feet.

"I don't . . . I can't believe it, but . . . just a minute. I'm going to check." And he left us, shutting the door softly so as not to wake the children. When he returned, he was muttering to himself. "That's the boat, all right, but I still don't believe it!"

"Something wrong?" Bill inquired.

"That's *my* boat! You took *my* boat."

"What?"

"Yessir. That's actually my boat. I towed it over there last week to be repaired. I didn't dare run it over because I was sure it would sink. It didn't . . . how did you get the motor to start?"

"We didn't," Bill replied. "We rowed."

"Rowed!" he exploded. "You mean you rowed that thing with all of you *in* it? You're all here, so I guess I've got to believe you, but I don't. And it didn't sink?"

"Not completely," I replied, "but it wasn't easy." He laughed. I was too tired to laugh.

"But you don't have to worry about getting it back where it belongs, because it's here."

When the taxi arrived, we carried the sleeping children down the rocky path to the dock, thanked our host, and left him still muttering his disbelief over his ancient skiff.

We spent what remained of the night at Basswood, Tom's island. In the morning he arranged to have the *Jackstraw* salvaged and took us home in an outboard. The wind was up, so it was a rough trip. He didn't suggest that I take the tiller, and neither did I. Entirely resigned, I took the bow seat and "balanced the boat" all the way back to Hay.

10

The Pontoon

The story of the islands is the story of the river, and to tell that is to speak of boats. They come in all shapes and sizes, from the classic, double-ended, cedar-strip St. Lawrence rowing skiff—than which there is no sweeter craft for the oarsman—to the great lumbering truck of the diesel-powered twin-screwed houseboat. Some of the big marinas in the area have the courage and insurance coverage to advertise the fact that they are willing, for a consideration, to put in the hands of any tourist one of these twenty-five-thousand-dollar combinations of steel and wall-to-wall carpeting. Such consideration paid, that person, unknown and unknowing, then has the right and privilege to go thumping from shoal to shoal at his own peril and that of other innocent boatsmen. Of course, the marina supplies the newly commissioned captain with all the necessaries, such as charts and so on.

"If you just remember to keep kind of a close eye on the chart . . . why er . . . why er . . . everything ought to be all right. No sweat, really."

The chart is indeed a marvelous invention, as is the broad, sturdy, seamanlike table in the captain's cabin on which the chart can be unrolled and studied to the heart's content. The wisely studious captain can learn everything he wishes to know from that chart as to the depth, say, of water on the lee side of Huckleberry and the rocks off Fort Wallace—it will tell him everything in clear, simple, diagrammatic terms that he can master in five minutes—everything except precisely where his craft actually is at any given moment in relation to those rocks. And that information can be—indeed, actually is—critical. More than a few skippers, experienced as well as inexperienced, have had their contemplation of their charts interrupted by a sharp jolt and the grinding of granite teeth along the hull. Those rocks speak in language hard to misunderstand and impossible to argue with.

But surely, you will say, *surely* there must be instruments that . . .

You are right. There are instruments, very refined, sophisticated, and expensive, that, when working properly, do not lie. But even the most expensive and superbly functioning instrument requires that a person read it, that he understand what he reads, and that, after he understands it, he act wisely in accordance with what he has read. The last freighter that tried to climb ashore stopped, fortunately for the cottager, with its bow hanging just over this fellow's roof—that freighter was half a mile out of the Seaway. Churning along as happily as if in the dead center of the channel, the engines of the ship deposited the two- . . . or was it four-? . . . million-dollar cargo of Christmas trees and toys bound for England almost in the living room of a fellow who was quietly minding his own business on his small island near the International Bridge. The freighter had instruments, too—more instruments than the average sailor knows what to do with. And so, presumably, do those other ships that run aground up there while plying their trade between the Great Lakes and European or Asian ports. In those waters, the last martini can turn out to have been an expensive drink.

Money—money—money! Must it always and forever be *money?* One hopes not, but still, when one has to work to get it, one is—that is, I am—overwhelmed by the courage of a person who, after plunking down maybe forty-thousand dollars for a cruiser, is crazy enough to climb aboard, turn on the engine, and take the bloody thing out where at every moment he risks total disaster. It's not for me. I'm just a tinhorn, a very smalltime gambler. My nickel-and-dime outboards are all I care to hazard on that beloved, uncertain, glorious, treacherous water.

Of course, even I have my dreams. At the bottom of that skinflint, penny-pinching heart of mine, a shadowy captain strides casually across his wide expanse of deck and marches to the stern without clambering over seats to where the big cooler sits, its ice rattling cheerily inside with each slow roll of the big hull. Oh yes! Down there is that ghostly captain who longs to stroll, unfettered, about the generous proportion of his ship as it steams along through the waves, instruments or no instruments, charts or no charts. In what heart does there not lurk such a shadowy figure, stretching forth from the gloom of dreams toward the mastery of a well-found craft? Also, like every dreamer, he longs to moor his fancy beside the dock of reality, there to translate his fictions into solid planking and rails.

Such a skipper lived in the heart of Hunnie's husband, Tom Murphy. A good and thoughtful man, husband, and father, Tom put a reasonable rein on his aspirations. No sleek racing sloop, however beautiful, would be practical for the father of eight. It had to be something portly, reliable, sensible, and it was—a deck twenty-six feet long and ten wide, surrounded, except for two gateways that could easily be closed off with chains, by a belt-high aluminum fence, and the whole riding on two huge steel, cigar-shaped pontoons. With a forty-horsepower West Bend clamped on its stern, he had a seagoing porch that was safe and comfortable for almost innumerable children, perfect for the odd Saturday with a crew of brave, beer-drinking fishermen, and suitable almost for a dance or clambake. Either *en famille* or with bands of lusty comrades, he plied the famous fishing spots of Oneida Lake north of Syracuse until he graduated to the St. Lawrence by becoming the proud owner of Basswood, a small island two miles off Fisher's Landing and about ten miles from Hay. Once there, his mighty and infinitely adaptable craft did yeoman's service in any number of vital ways.

But, in the fullness of time, princes, even queens, grow old and frail. As with them, so, alas, even with the pontoon boat, a decline that had special poignance for me. I had, after all, done Tom the honor of sinking his *Jackstraw,* so I experienced some guilt about his transportation problems. The summer must have been pushing August when I noticed the pontoon boat still in its winter berth, dragged halfway out of the water and firmly chained to a hefty tree. I asked Tom what the trouble was.

"Oh," he said, thumping the still bright and sturdy rail, "pontoons. There's a leak somewhere in the darn things, and it's practically impossible to find out where. They're steel, you know, not aluminum, and they've probably rusted out. Shame. It was a great boat, but it doesn't owe me anything, not after all these years."

It was at this point that the shadowy captain in my own heart waggled the empty glove of a hand to attract my attention. Instantly I knew what was going through the depraved mind of that scoundrel. No shame—absolutely *none!* If you can believe it, what he was proposing was to offer Tom some piddling figure for "this . . . this *wreck!*" as he would put it; then to hustle it somehow back to Hay Island where, with the help of the ingenious Durrell, he would make it seaworthy once more. That accomplished, he intended to go haring up and down the river, glad to be the envy of all who saw. Oh! the degradation of which the dreamer in me was capable! He was shamelessly anxious to bring to birth his own glory on the disappointment of another, of a brother-in-law, of a friend, even the disappointment of a friend who lived close enough to be a spectator of his triumph. In-*credible.*

But with one fierce gesture I put that wretch in his place and shut him up—for almost two years. Two years during which the pontoon boat lay forlornly chained to its tree, accepting its burden of autumn leaves and winter snows. Then one afternoon, my guard down as I fingered a cold draught with Durrell in the International, he spoke his black thoughts in soft and wheedling accents.

"Now, there's a thought," said Durrell thoughtfully. "There *is* a thought. Because there's nothing, no machine . . . that can't be fixed."

That was encouragement enough for me, and we spent the rest of the afternoon trying to figure out how to fix this one. Though we could come up with nothing promising, I still had faith in Durrell's ability, joined as it was to his Pennsylvania Dutch determination. This provided the gall needed to broach the subject with Tom on our next visit to Basswood.

"What do you say, Tom? How about a hundred dollars for her as she lies?" He digested the offer, fathomed its infamy, and utterly outflanked and unmanned me by a thrust of his automatic generosity.

"Tell you what I'll do," he said calmly. "To accept that would be an insult, an insult to the boat, I mean. Instead, I'll give it to you." He reached for his wallet, fished around in it for a minute, and then handed me a plastic card. "Here's the registration. You've got to sign it, that's all. Hand me your glass there. It looks kind of low." He took it and went on as he poured, "Now, about the motor . . . I'll just lend you the motor. We might need that sometime, so it will only be on loan. Okay? Understand?" I understood that, plus quite a bit more. "But if you get those leaks fixed and the motor working, maybe you'd come over and give us a ride."

Ever sensitive to the feelings of others and inexhaustibly kind, Hunnie came to my rescue as I sat holding the registration. "Go ahead, Rob," she urged eagerly, "go ahead! If you don't, that boat'll just sit there. We'll never do anything with it, and it's no good to anybody sitting there on the shore, so you go right ahead!" She laughed. "Of course, I don't know how you *can* go ahead. It'll just sink if you try to move it. Have you got any ideas about how to go about it?" I said I had done some thinking, but not enough, not yet.

"It's all very well to feel bad," said Durrell the next afternoon, "and I can understand that, but let's face it! Be kind of nice to have that thing floating around here, wouldn't it? Besides . . . what the hell! Tom knows very well he's only got to say the word and he can have it any time he wants and for as long as he likes—that is, if we can fix it up."

"But you said," I reminded him, thumping the dock where we sat, "you said there was nothing that couldn't be fixed."

"Yeah, but I say lots of things," he laughed. "Just because I say them doesn't mean you've got to listen. Still," he cleared his throat, "we can afford to be serious now you've got the registration, if you know what I mean. Let's lay this thing out and approach it logically. Nobody's going to make pontoons that big without putting in plugs for draining them." He hadn't examined the pontoons, but I had.

"Oh, there are plugs, all right, one on the very top of each pontoon at the back to let in air, and another at the bottom of the back end to let out water. The only trouble is that somewhere along the line someone wanted to save fifty cents on plugs, and they put in cast iron instead of brass. Those plugs are rusted in so solid it's like they were welded there. Somebody's tried to unscrew them, too, and they chewed the corners off all of them, so there's nothing you could even get a bite on even with a pipe wrench."

"Okay. So that's no good. Now, let's have a look at the picture. We've got three problems: One, to drain out the water in the pontoons; two, to find the leaks where the water gets in; and three, to close up all those holes so the water doesn't get in so good once't any more," he said, putting on his Pennsylvania Dutch accent. "You agree?" I agreed.

"The first is simple," he continued. "Drill a hole in the back of each pontoon at the bottom, and Isaac Newton will take it from there. Now, the second's harder, but let's try this on for size. If we can drill a big enough hole down through those plugs on top, big enough to tap the holes with thread for a quarter-inch pipe—if we can do that, we could screw in a quarter-inch nipple—I think most fittings on air compressors are quarter-inch—and then we could hook up a compressor and pump air into the pontoons."

"Wait a minute. Wait! What would we want to do that for?"

"Bubbles," he explained. "You see, the air under pressure will come scooching out through the holes—those leaks we can't find—and make bubbles in this dishwashing stuff we've painted the pontoons with. Well, we'll be able to see those bubbles, mark the holes, then seal them up with maybe some of this new metal putty."

"Metal putty! What's that?"

"Never mind about that. It's some new stuff—I've got some in my toolbox—that you mix up, and chemical action hardens it till it's as strong as metal. Anyway, I think that would do the job. How does it sound to you?"

"Sounds marvelous," I said; "just great, but do you really think it'll work?"

"Says so right here," he said, tapping his forehead with a finger. "So we've solved all the problems, but we still can't take her out for a cruise. Talk's fun, even if it is cheap. Bet that'll do the job, though."

Lots of problems can be solved sitting on a sunny dock, but there's still the "doings," as Auntie Em would say, "doings," like just laying hands on an air compressor, for example. Not everyone has an air compressor tucked away behind the detergent bottle under his kitchen sink.

"What the hell would I be doing with a thing like that?" came Jack Stuart's voice over the phone. "I haven't got one, and what's more, if I had one, I wouldn't lend it to you . . . anyways, not until you give me back that nice big jack of mine. How many years have you had that?"

"I didn't really expect you to have a compressor," I said. "Just bring one with you tomorrow night when you come over for poker. If you do, you can have your jack back." He laughed and said he would try, but that he couldn't think of anyone offhand, and had I tried the hardware stores. I said I hadn't, but that I would.

"And if there's any room at the game, how about giving Duffy a call," he went on. "I haven't had any of that little bugger's money since he got married."

I tried the hardware stores without success, and Fred Lawton at Mitchell & Wilson, the lumber company, just laughed.

"What are you up to now that you need one of those things over there?"

"Too complicated to explain," I said and hung up. Then I called Duffy, and Duffy agreed to bring some of his money over for Jack to look at if I would let him, Duffy, give me an estimate on painting the place.

"What does an estimate cost?" I asked.

"Estimate's free," he replied.

"You're on then," I said, "but don't bother bringing a paint brush."

Around ten the next night, Duffy had an ace showing in a hand of two-card high-low, and he was raising dimes every chance he got. Durrell was groaning and throwing in his money, and Jack kept pushing what looked like a pair of kings. I folded my hand and turned to munching a pretzel, drinking my beer, and making snide comments at the players. In the middle of a semi-witty crack I was struck speechless by an actual thought.

"Duffy!" I cried, "exactly how do you paint a big . . . I mean a really big building?"

"Up and down," was his reply. "And ten better." He tossed in another dime.

"No, damn it! I mean do you use a *brush* or what?"

"Sprayer. I'm in."

"Then I bet you use compressed air."

"Till I find something cheaper," he replied.

"Then, by God, you've got a *compressor!*"

"Yep. You going high or low, Durrell?"

"Low," said Durrell.

"But you're the guy I've been *looking* for. How'd you like to lend it to me?"

"Lend, *nothing!*" interrupted Jack. "He told me only yesterday he was willing to *rent* one, Duffy. You stick it to him, Duff."

"Yeah, but . . . well, that was yesterday," I said. "How about it, Duff?"

"Jesus!" exploded Jack, throwing down his cards as Duffy flipped his ace in the hole. Duffy reached out and began counting his half of the pot, and Durrell worked on his half. "And you just keep quiet, Russell," Jack continued, "until the man gets his money. A pair of *aces!*" he mourned to nobody in particular but everybody in general. "He had 'em all the time."

When Duffy had finished counting, he said: "You want a compressor?"

"I sure do," I answered.

"Well, then, this is your lucky day because my goddamn men are on strike so nobody is using it."

"What do you want for it?" I asked.

"Jesus Christ," he roared, "I'm not selling it! I want it back in working order, that's what I want. Whose deal?"

"Can we pick it up in the morning?" put in Durrell.

"I'll leave it on the front porch," Duffy said.

"That's enough of this chitchat!" complained Jack. *"Deal!"*

"Some guys are just born lucky," commented Durrell as we loaded the compressor into his station wagon next morning. "If you arranged that painters' strike, I hope you remember how you did it because we might need this thing again sometime."

Two hours later, at Basswood, a healthy stream was pouring out the back end of the starboard pontoon. The port pontoon had turned out to be bone dry. As instructed, I had drilled holes through the chewed-up plugs on top of each and was tapping them with a quarter-inch thread while Durrell was talking the compressor into operation. Sundry small Murphys were excitedly sloshing detergent over the starboard pontoon. Tom, Hunnie, and Elisabeth were watching with mounting excitement. My tapping finished, I screwed in the short length of quarter-inch pipe as Durrell coaxed the compressor into action.

"Here we go," he said cheerfully, hooking up the hose from the compressor with its quarter-inch fitting to the short pipe. (Back at the International, he had said casually, "most compressors have quarter-inch fittings." How any college professor could toss off a remark like that and actually be right is beyond me, but he had, and he was.)

As the pressure inside the pontoon grew, the stream of water from the hole in the back end arched out like that from a garden hose.

"You know what?" Durrell said, licking dry lips, "that water's beginning to get to me. What time is it anyway?"

"It's not too early," laughed Tom. "Hey, Mary Ellen, would you run up and bring us a few beers so that we can wait? This may take a little while." And Mary Ellen ran off toward the house.

"Wait, *nothing!*" cried Durrell, diving toward the bow; "there's one of our leaks already." Chorus of cheers.

"Will you look at those bubbles!" Tom admired. "And there! There's another one up on top, right by that weld!"

"Dad! Hey, Dad!" cried Tommy Murphy eagerly, "can we paint the other pontoon with this stuff?"

"Don't have to," answered his father. "That one doesn't leak."

"Yeah, but can we paint it anyway?"

"Sure. I guess so."

"Gee, thanks. Thanks. Come on, Jimmy," and they went to work with their brushes.

After marking the places to be patched, we settled down to wait for the pontoon to be pumped dry.

"But those holes," Tom mused, "those holes you drilled in the back end . . . how are you going to seal them off? But then, I don't suppose you really want to seal them off permanently because you may need to drain the pontoons again sometime."

"Boiler patches," was Durrell's answer. "Great little gadgets, boiler patches. Screw them up and they seal the leak. Unscrew them, and there's your hole for draining. Pretty clever things, really."

"Sounds great," said Tom, "but where are you going to get anything like that around here? We sure haven't got any . . . not that I know of, anyway."

"No problem," Durrell replied. "I picked up a few the other day and brought them along in my toolbox."

"Is there anything you *haven't* got in that toolbox?" Hunnie asked.

"Haven't got a compressor," admitted Durrell, "but now I know where I can get one, and that's the next best thing."

For the four quiet happy hours it took the pontoon to drain, we quietly and happily absorbed. When at last the pontoon was ready, we were not quite. But then, when is man ever actually ready, except in dreams?

With boiler patches and metal putty, we plugged the holes. Then another peaceful hour while the putty hardened. Then the launching. All hands, both large and small, plus many feet, were called into service to dislodge the mighty craft from dry dock. The boat, or, more properly, porch, perhaps patio, groaned its painful way down into the river.

"Sorry, but we're fresh out of champagne," said Hunnie, snatching up the empty plastic detergent container, "but this will do," and she beat it on the prow of the nearest pontoon. "Darn thing won't break!" she exclaimed, "but anyway, consider yourself christened, boat!"

"It floats!" I cried. "The darn thing actually *floats!*" as the back of a pontoon bobbed gently beside me in the thigh-deep water.

"Did you expect it to sink or something?" inquired Durrell.

"Come on back in, Rob," called Tom happily. "I'm going to need a hand with that motor."

Another half hour's heaving, clamping on, and Durrell's tinkering with the ancient West Bend, and she popped, barked, and finally consented. Turning from the running engine to the wheel, Durrell issued an invitation for everyone to come aboard for a test cruise.

"I'd come," said Tom, sinking down onto a rock, "I really would, except I just plain don't believe what I'm seeing . . . not yet, I don't."

It was almost dark when our convoy rounded the head of Hay Island and we moored the pontoon boat by the boathouse.

"So," commented Durrell, leaning against the rail, "now we've got a real workboat. This thing," he patted the rail, "can carry almost any amount of stuff—this will solve a lot of problems, you know that? Now we can get down to some really *big* jobs."

"How about a day off, please, sir?" I requested. "Anyway, time to get Duffy's compressor back to him."

He considered the request. "Suppose so," he said grudgingly; then added, "Just so long as we don't go getting tangled up over there in town and start enjoying ourselves!"

The patches held and the boat proved useful and pleasurable in ways too numerous to go into. The only thing about it that still bothered me was finding how it could be helpful to Tom and Hunnie over on Basswood.

"When we need it," they kept saying, "we'll let you know. Don't worry! But at the moment we're concerned about our boathouse." They had a problem there, all right. It was an enormous structure, and fifty years of being admired without being repaired had left it leaning dangerously—so dangerously that it was downright unsafe to go anywhere near it because it could, at any moment, collapse. A match was really the only solution—a match and a little gasoline. I have to admit they did invite us to the party, but they held it too late in the fall for us to come. We weren't the only absentees, either. With his customary forethought, Tom had invited a nearby fire department, invited them well in advance, and got the guarantee of their presence and expert supervision. On the strength of this, he had, as was his habit, laid in ample provisions, enough to gladden the hearts of several fire departments, but they didn't show. The Coast Guard, whom he had forgotten to invite, did show in their cutter to investigate the blaze and only narrowly escaped both running aground in the unfamiliar waters and being themselves set alight as they struggled to maneuver into deeper water.

A grand occasion it must certainly have been, from all reports, but, like the fall of Troy, the sack of Rome, and the rape of Lucrece, it was one of those historic events a fellow has to get used to the idea of having missed. Like most such events, this one, too, left its marks. These consisted of the charred remains of what had once been ample docking. Of course, the plans hadn't included dock burning, but fire has a contagious way about it. The Murphys were so relieved by the fact that the house hadn't gone up as well that they had no time to worry about the dock. Still, enough docking space had been lost to cause them serious inconvenience and to cause visitors like us even more.

"You know what," I lectured one afternoon on their porch, "the only thing to do is to clear away those charred timbers and build yourselves a decent dock."

"Yeah," Tom agreed sleepily, "that's the only thing."

"Cedar," I continued, "you've got to use cedar!" He agreed again, but added that he had tried the nearby lumber companies. They neither had any cedar nor would they deliver to islands. I considered this for a while.

"Tell you what," I offered. "I've got a pretty decent workboat, and I know where I can get all the cedar you need." He sat up. "And maybe we can talk Durrell into supervising the job."

"Oh! . . . Oh, I couldn't do that. Poor Durrell! He gets put upon by everybody and I couldn't . . . I just couldn't . . ." but he was up and pacing the porch.

"That's the price he has to pay for being a genius," I said. "And if he knew you really wanted to do it, I bet he'd offer."

"And you think you can get the cedar?"

"Arvilla?" called Durrell, "have we got anything on for the next couple of days?" She said they hadn't, and wanted to know why he had asked. "Well, Bob here is making me a pretty attractive offer. Promotion included. *Supervisor!* How does that sound? Bet you didn't know when you married me that I was going to go right up the ladder."

"Those were the days before you even had your Ph.D.," she said. "What's the job?"

"Dock over at Basswood," he explained.

"Fine," she said, "as long as I can come."

"Good," he said getting up. "Where's Norman? Medievalists can hammer and saw almost as well as ordinary people."

The pontoons completely submerged with the load of cedar timbers and topping, the ripples playing at the base of the rail, we crept in to tie up, unload, and begin at Basswood. Our supervisor had the whole outfit going like an anthill in short order. Those too small to either swing a hammer or push a saw were pressed into a refreshment detail, and for three days the little island fairly hummed and sang. When the song died, Basswood sported a new hundred-foot section of dock five feet wide and all cedar.

"Next thing," sighed Tom, relaxing in his wicker chair on the porch, "the next thing I've got to do is to rip off this section of porch railing."

"What for?" asked Arvilla with concern.

"Because from here . . . from right here where I'm sitting," Tom went on, "this railing blocks my view of that last, beautiful twenty feet of dock, and I mean to spend a good bit of time sitting right here admiring that job."

"You're going to sprout roots," said Durrell, "if you mean to sit there watching till that dock moves."

"Yeah," said Hunnie with dreamy admiration. "But you know what!" she added, "that business about going through those yellow pages to find the right outfit to do a job . . . that's a lot of bologna! For any job, repeat, *any* . . . all you really need is a boatload of professors!"

"Long as Durrell is one of them," I said.

"Right!" they chorused. Then, a long, thoughtful silence that Tom finally broke with, "But this . . . this sort of thing . . . why doesn't this sort of thing happen other places? Doesn't happen in Syracuse, anyway."

"I don't know about Syracuse," said Durrell, "but in Lancaster, they won't let you run a hundred-foot section of dock out into the street, not even if it *is* cedar."

"And," I put in, "over there on the mainland, nobody I know passes around pontoon boats. In fact, over there there aren't any boats at all. There's nothing . . . really nothing for captains to do over there . . . nothing but dream."

11

Middle Age

Middle age is not definable in years but in how we recognize and face a basic question. Since life is for action, what am I to do with my time? To what work will I give myself and for whom? Deciding these issues is the task of our youth. Deeply absorbed in them, we alternately drift, leap, hesitate, and plunge toward what seem answers. For good or ill, right or wrong, most of us settle these questions in our twenties and set about the business of building a future in jobs and families, and for ten, fifteen, perhaps twenty years, this is enough to occupy us. But then, slowly, slowly, and perhaps especially to those whom we call successful, comes the question: "Is this all? All you are going to do with your time here? You have earned a measure of respect. You have had your children, fed them, cared for them, done your best, and now they no longer need you as they once did. Man? Woman? Are you simply going on in this same way to the end? Till death? Is the important action of your life finished? Time grows short. Have you anything different to say, to do, to be? Are you finished? Your account is closing . . . closing! If there are any different sorts of entries you want to make in it, you had best make them soon, make then *now!* What's your pleasure? Or are we to expect the same sort of thing from you until the end?"

Whatever you have achieved—however much, however little—the question presents itself with the same urgency: "Is that *all?*"

To face this question is to be middle-aged.

Not many among us can say with an untroubled heart, "Yes, that's it. That's all I am. All I have been I shall simply continue to be to the end. There are no surprises in me any more. As nearly as I can make it, my future will be a repetition of my past." Which of us is satisfied to go on like this, plowing the same furrow till it becomes a rut, a ditch, and then a grave? Whatever we do, we must face the question and decide.

Some reject its challenge by turning to the bottle, TV, bridge, travel; and some accept like so many old champions who come out of retirement and step back into the ring on unsteady legs. These divorce, go on diets, jog, remarry, try the ministry—everyone determined to catch what cannot be caught, that clean page on which to spell out the story of one's life. But, clean page or no, we must try to pretend it is clean because we mean to begin again and this time . . . this time we shall do it right! And why not—considering the alternative?

This central question gradually began to present itself to us—not clearly, not urgently, but giving rise, nonetheless, to vague unrest as we realized our lives had fallen into a pattern. In each nine-month sojourn on the mainland, I became submerged in academia. Campus politics flared over this issue or that.

"But, Bob! Don't you think it's outrageous that . . ."

Well, I thought so or I didn't think so. Then someone—too often I—wrote a report on the matter and circulated it; the issue brought to a vote, my side either won or lost, and things died down. There were courses to be prepared for and taught, examinations and papers to be read. From eight in the morning till eleven at night the days were full. But, as my position at the college became more secure, the imagination stretched forward to envisage thirty years of the same.

At home, Elisabeth did the laundry, got the meals, cleaned the house, read to me, and taught part time at every opportunity. Richard and Mark took their sandwiches each morning and went off to school. As time passed, even James and, at long last, Miranda followed them and left Elisabeth to the silent, but insistent demands of an empty house. The corridor of her future stretched out into decades of emptier and emptier houses.

The summers were a relief for us all but, in comparison with the long dark winters, they fled like dreams. And even the dreams seemed not quite so full as they had in those first years on the island. The dock was built, the porch fixed, the roof hardly leaked, the grass was annually laid low by our power mower (shades of Mrs. Blanchley-Hoyt), a Sears electric pump mounted on a cement platform down by the river quietly and efficiently supplied us with more water than we could use and with more pressure than we had in Lancaster. Under the kitchen sink, wonder of wonders, squatted a five-gallon hot-water heater. Though I had lain under that sink on my back, the puddles beneath me richly fed by showers from poor connections overhead, tightening this clamp, then that—that job had been done, and hot water obediently flowed from the tap marked "hot." Also, the icebox had been replaced by a dependable, though ancient, coil-crowned G.E. refrigerator that purred contentedly beside the bottle-gas stove. The disposal of that icebox had, it's true, given trouble.

To celebrate the plug-in of the G.E., I had, with the help of medieval Norman, dragged the obsolete box down one dark night and laid it across the bow of the now old fiberglass, which had, as advertised, proven to be stronger than steel, and which had not shattered, or sunk, or anything. Our intent was darker than the night itself, for we meant to scuttle the box in the deepest part of the channel. Arrived at a likely spot, we heaved and swore, swore and heaved. Suddenly, the load shifting, the boat heeled far over and the St. Lawrence streamed in for an instant as the mighty box found with a splash its hoped-for resting place. But it wasn't. When our cheers died away on the night air, the ungainly hulk of the box bobbed gently on the surface beside us.

"Goddamn thing won't *sink!*" Norman howled. "It floats."

"It can't!" I prayed. "It's just that the doors are shut. It'll sink . . . it's got to sink when it's full of water."

"Try again," he encouraged. "The doors *are* open. Let's just get the hell out of here!"

"We can't, for God's sake! The bloody thing's a navigational hazard. Some boat comes haring along here and hits that . . . we're responsible!"

"You mean you're responsible," he corrected. "Not my goddamn icebox. Still, you're right. What the hell we supposed to do? We can't get the goddamn thing back in the boat!"

Leaning far over the side, the water past my elbows, I secured a line around one of the gently waving doors. Like a sluggish, surly monster, it heaved, rolled, and swung behind us as we towed it slowly back to shallow water. Next morning I dragged it ashore and gradually dismantled it, taking the separate pieces to the refuse bins at the town dock.

Like other important island jobs, this one had been a lot of hard work spiced with a strong dose of the unexpected, and it was something to laugh about on winter nights. But it, like other and bigger jobs, was done. And while the sun shone just as warmly on our front steps, and the catbirds sang, and the winds blew as they had sung and blown, and the Olde English fruitcake, reduced for quick sale from the A&P, still contained in its stodge its soggy surprise of raisins, something had departed. We were a good deal more comfortable; so the vague, continuous sense of mild disappointment seemed hardly fair. Of course, the hymn singers up at the head of the island, those hundred virginal voices raised each morning at eight in praise of the Creator—they didn't help either.

Hay Island is shaped something like a fat-handled hatchet running northeast to southwest, with the head or blade at the southwestern end. This sixteen-acre blade had once been the estate of someone named Lewis. With its cliffs, bays, rocks, and trees, it was certainly the loveliest part of Hay or, for that matter, of almost any of the islands. It had been on the market for some astronomical price during our first year. Unfortunately for us peace-loving islanders, it was bought by two or three public-spirited service clubs in Kingston and turned into the Mohawk Girls Camp. Every Saturday in July and August, a tour boat from the Gananoque Boat Line would flounder from town to the huge Mohawk boathouse, laden with another hundred twelve-year-old girls for their week of fresh air and sunshine.

Every night the underbrush twitched with scouting parties of amorous young braves from town, and the river boiled with adolescent outboardsmen "showing off" or trying to sneak in for a secret landing on the camp's mile of shoreline with the intention of counseling the counselors on matters of love. The only male legitimately on the premises was the caretaker—Whiz, by name. The name was appropriate in as much as that is what he would have to have been, but wasn't, if he were to protect the virtue of all of his charges. On that job, Clark Kent would have had to work overtime.

If the nights were bad, the mornings were worse. Still chockfull of zing from their nocturnal scrambles, the girls raised shrill voices in chorus through unnumbered verses of *Holy, Holy, Holy* at their eight-o'clock hymn sing. Their vigor brought no joy to the heart of a solitary fisherman on the end of his dock, idly casting out toward the channel with middle age slackening the tension of his bones.

"You're exaggerating!" said Elisabeth one morning when I came up from the dock with my grousings. "Besides, it's jolly nice for them. Why shouldn't they have a vacation on the river too?"

"They bother me," I countered sullenly.

"Too bad," was her unfeeling reply.

"Worse than that," I observed.

But it wasn't "too bad" in the summer of '66. The first Saturday in July came and went quietly. No top-heavy tour boat chugged out from town to deposit its cargo at the Mohawk boathouse. The following Saturday was equally peaceful. In fact, the whole summer swam leisurely by with the lounging Whiz as the only occupant of the camp, and he was happily not given to hymn singing of a morning. With no maidens to guard or invaders to chase, he casually turned his attention to the wildlife, and an occasional shotgun blast was his only notable contribution to the symphony of summer sounds.

"Been a quiet summer," I observed one afternoon to Cliff Carpenter, whose property adjoined the camp. As editor of a Rochester newspaper, Cliff was shoulder deep in the turbulence of a large American city, so he especially prized week-end island peace.

"Great, just great, but have you heard the news?" he answered.

"I never listen."

"It's people like you that make it hard for people like me to earn a living," he complained, "but I'm big-hearted. I'll forgive you if you'll just listen."

"Must be bad news or you wouldn't be so eager," I said.

"Newspaper training," he apologized. "The point is that those beautiful, well-organized moderately peaceful campers were an annoyance to you . . . to us, but that was a Sunday-school picnic compared to what may happen here. The grapevine says that Mohawk is up for sale, and the government is interested."

"What's the government interested for?"

"A public park! What else? The government has this notion that the people in Canada need recreation and land on which to recreate. Pleasure isn't only for you rich Americans, you know."

"Us rich Americans," I corrected.

"Anyway, the government feels that Canadians deserve places where they can water-ski . . . starting at six every morning after their all-night parties where they've been playing their accordions and trumpets. You think those poor, dear little virgins were noisy . . ."

"But they've got islands . . . whole islands . . . lots of them," I objected.

"Yes, and they want more. Can you imagine what it will be like? Loud parties every night and *all* night, and us—you and me—not invited?"

"How far is Hudson Bay?" I asked.

"Forget it. I looked at the map." He paused.

"Don't you know anyone in Ottawa?" I suggested. "Can't you threaten an exposé of their sex life or something unless they lay off?"

"Canadians are universally virtuous," was his sad reply. "Must be the climate or something."

"You're a newspaperman," I urged. "I thought you guys were schooled in the ways of the world! There's got to be an angle somewhere!"

"I am; we are; and there is," he congratulated. "Every problem has its solution."

"Well?"

"Not even complicated, either. Buy Mohawk yourself. That would solve the whole business."

"It might solve the problem," I agreed, "but it would *dis*solve me. Why don't *you* buy it?"

"It's only money," he evaded.

"But it's *mine,*" I objected. "Or, rather, it isn't mine. Where am I going to get . . ."

"That's the whole point, Russell," he explained patiently. "It *isn't* your money, so why not, for heaven's sake? Almost any bank would be delighted to buy Mohawk for you. That's what they're in business for. Any bank would be proud to help you out. You're a first-rate risk. With my heart attacks, they wouldn't lend me the time of day. But *you!*" He slapped me on the thigh. "Why, if I had it, I'd lend it to you myself."

"Thanks."

"They want fifty, but they'll take thirty. Offer them twenty."

"And where would I be if they accepted?" I objected.

"You'd be sitting on the end of this island, the luckiest man in the world," he said, getting up to leave. "Just think it over, but not for long. Remember what this island would be like if Hay were a public park. Just think about that, and invite me to your housewarming." Having darkened the day, he left cheerily.

"What does he know about it anyway?" Elisabeth said sullenly that evening. "Fifty thousand, but they'd take thirty. How does he know that?"

"Grapevine," I answered.

"Grapevine, grapevine!" she said contemptuously. "What does he know about grapevines?"

"Be great," chimed in Richie. "That big boat with the sixty! Be great for water-skiing!"

"You can get a big boat with a sixty horsepower motor for a lot less than fifty thousand dollars," she said dryly.

"Thirty," I corrected.

"Well then, what are we waiting for?" Richie returned.

"Be quiet," said his mother.

Next afternoon we took a walk up over the cliffs and around the deserted camp, something we had never done while the camp was in session. Leaning against a tree and thoughtfully chewing a toothpick, Whiz glared at us dutifully as we approached.

"Hi, Whiz," I called cheerfully. "Quiet summer, eh?"

"Yep."

"Mind if we just walk around?" Elisabeth asked.

"We don't encourage that," he said gloomily.

"Oh no. Certainly not. But, Whiz, we've heard rumors that the camp is up for sale." He made no answer. "Heard anything about it?"

"Could be," he admitted.

"But you don't know for sure."

"Nope."

"So what if it is so beautiful!" Elisabeth said, angrily thumping the top step where we were having tea after our return. "What if it is? We don't even know if it's for sale. Even if it is, we don't have any money, and what's more," with another thump, *"this* is ours, and I like this just fine. So what good did all that tramping around do us?"

"Didn't you enjoy the walk?" I asked.

"Yes . . . no. Well . . . thinking about the camp . . . it's all crazy. . . . Anyway, we're just nicely settled. Just nicely settled, and you want to go rousting us up again! Just like you!"

"It's been eleven years," I observed. "Call that rousting?"

"You know very well what I mean."

"Then forget it. Forget the whole thing."

"Oh!" she exploded with annoyance, "I've got to do the washing," and she marched off. Lazily, I finished my tea, then went out to fiddle in the toolshed where I could hear her stamping around in the kitchen till dinnertime.

At the table, she said: "What did you think of the camp, Richard?"

"Vroom-vroom!" Mark interrupted.

"Quiet!" she exclaimed. "If you're so interested in water-skiing, why don't you make some rich friends?"

"Don't need them," said Richie. "We've got you, Mom."

"And I've got your father," she complained.

"We've got a marvelous fifteen horsepower engine that works on the back of a monstrous fiberglass boat," I said. "Damn it! When I was a kid, I thought it was the greatest thing in the world to rent a rowboat for the afternoon for fifty cents on a trickle of water called the Chenango, and you kids . . ." They howled.

"We know that one by heart, Dad," Rich patronized.

"Yeah, Dad, we know that one by heart," Mark chimed in.

"Your father has to make his little speeches," said Elisabeth. "Leave him alone and let him finish."

"Well, damn it! It's *true!* I never even dreamed of having my own rowboat. . . ."

"I'm going to have a cruiser when I grow up," put in James.

"I bet he will, too," laughed Richie.

"And I'm not going to tell my wife I have it, either," he went on.

"Why not?" asked Elisabeth.

"So I can go off by myself on weekends," he explained.

"Sixty horsepower motors and cruisers!" I growled.

"I don't call them very much," Elisabeth jabbed, "and here you are wanting to buy a whole great girls' camp!"

"I didn't bring the subject up," I objected. "You did."

Delighted cries of "Hurray, Mom! Hurray, Dad! Go to it!" from the children.

"I haven't said a word," I declared firmly.

"But you've been thinking about it," she countered.

"And so have *you,*" I snapped.

"Then why shouldn't the kids think about their sixty horsepowers? We're the affluent generation, aren't we?"

"But *you* were the one said we didn't have any money!" I said.

"Well, how can I not think about it when it's so beautiful? What's the matter with that?"

"If that's the case, then we'll buy it," I said expansively.

"Yeah."

"Yeah!"

"Whoopie! Offer him twenty, Dad!"

"Yeah. Go on. Offer that guy in Kingston twenty!"

The next morning, I did.

The phone trembling in my hand, I hemmed and hawed to Mr. Birchard at the other end of the line. The children rooted to their chairs, Elisabeth marching vigorously around the room, stopping on her way past me to snarl: "Never mind all that! Make him the *offer.*" Turning to lean my back against the wall, I summoned the courage to mention the figure. It was followed by a long silence.

"Did he hang up?" Elisabeth whispered.

I shook my head. Then: "Yes. Yes, of course, Mr. Birchard. I'd be glad to. Good-by," and I hung up.

"What would you be glad to?" everybody wanted to know.

"To put it in writing," I said nervously.

"But you can't do that!" Elisabeth cried, taking my arm.

"What do you mean, he can't?" roared Richie. "He's got to. How can we buy it if he doesn't put it in writing?"

"But *writing,*" she quavered. "What would we ever do if they? . . ."

"They won't," I said. "At least, I hope they won't."

When my nerves had relaxed, I stirred them up again by typing the letter and actually mentioning "twenty thousand." I signed it, sealed it, and went down to the dock to put it in our mailbox.

Each morning that followed, as I went down to check the mailbox for an answer, the dock I walked upon seemed farther beneath my feet than my hammering of the nails had brought those planks and stringers. As I stood upon them, part of me seemed to be not quite standing on them, not quite in touch, not completely rooted in the familiar sense of possession, of owning through labor and being myself owned by what I had done. The cement was as solid as ever. The six-by-six timbers and the two-inch topping—all cedar—would last almost forever. The foundations had been drilled and lag-bolted into solid rock. They would never move. But I had moved, or part of me had, or, at least, was beginning to, beginning to leave what I had thought I could never leave. Could it be possible that I actually intended this, intended leaving? Was I actually arranging things so as no longer to be able to answer Elisabeth's call to come up from the river and sit down on the top step to drink my tea and listen to all the familiar sounds and feel the same sun and wind? Could I be meaning to leave this? Willingly? And for what?

For what? Surely not for a vast new place whose rocks I did not know, a shoreline so long that I could never learn it, buildings (how many were there?) I could not even find my way through? And for this privilege of being perpetually confused, lost, dependent upon the help of others to get me even from one part of my own property to another—for this privilege was I seriously proposing to put myself in debt for a decade or two? And that toilet! I understood that toilet. After long hours of apprenticeship, I had become its master. I could fix it and, after I had fixed it, it would actually work. It was a sort of gloomy, unreliable, but nevertheless, close friend. What would the new one—no, not one, maybe a dozen—what would they be like? Strangers! The whole pack of them! It would be an entirely new life! I must be mad! Those years of happiness must have imperceptibly swung me completely around the bend. That must be it. Otherwise how could I be proposing this?

Was it that force which so commands, that force which drives the sleepy and reluctant traveler on through the blizzard toward some warm, snug cabin built of his own dreamings? Not love of life, surely, because life had been possible, even comfortable at the previous stopping place. If we genuinely fear death, we would not risk moving; yet leave we do—leave, and leave, and leave. First, the womb, that perfect place of peace and plenty, our Eden; and for the cold, empty crib that gradually we come to know and accept as home. Then its pleasant, protective rails become the bars of a prison, and we cannot rest until we scale them, only to crash on a hard, indifferent floor in the vacant room. Though kind parents return us to the security of the crib, next day we struggle again over that same calamitous hurdle to crash down again and again till that no longer hurts. And we repeat the same process unnumbered times, learning to become comfortable in our nursery, only to reject all that by slipping through the door and falling downstairs. Mastering the house, it becomes our home; then out the door into the yard; out of the yard to our block, then our school, our city, our state, our nation—until, triumphant at last, we blast off in a silver womb from the earth itself. Are we searching? Or are we rejecting?

When we are young, the excitement of the unknown draws us forth from the cradle of the familiar, and we peer eagerly into the mystery of the future. In middle age, equally driven, we move more cautiously, more slowly, and with growing sadness in the knowledge that, though what we are leaving is good, it is not enough to hold us. So, over the years, our eyes turn from the future toward the past until we back with increasing care into a world that will eventually trip us, and we shall go down, not to rise again. But, even knowing all this, we cannot not go.

The morning Mr. Birchard's answer came was full of relief and disappointment.

"Then that," said Elisabeth firmly, "is that."

And so it was till one evening after dinner that autumn, when I began: "The question is . . . the question is, how much is money worth?"

"Augh! You and your mysticism!"

"It's not mysticism!" I objected, "and anyway, it's true. It's very simple. Just what's money worth, that's all?"

"The real question," she countered, "is how much Mohawk's worth and whether you want it that badly. They're the only questions."

"Me? I? I want it? How could *I* possibly want it? I've got no notion of what it's even like, for God's sake! How can I know whether I want it?"

"You've been around it same as me," she said.

"The only difference is that you saw it. I mean, how would you like to be trundled blindfolded over rocks and through the woods for half an hour and then asked if you liked the place enough to put out thirty-thousand dollars for it?" She made no answer. "Oh, I could get to know . . . well, parts of it, but it would take years, and we haven't got years to decide. The question is whether *you* want it." She thought for a while.

"It is fantastically beautiful," she said. "That's important, but not as important as whether we would be happy there, and how can we know that? I mean, we *have* been happy in the old place, but we didn't know before we bought it that we would be. I don't really know. I really don't." After a pause, "But it is so lovely! . . . those cliffs! . . . the bays! and those silvery rocks! and all those trees! Wonderful trees." Another pause. "Let's ask the children what they think. After all, it would be even more of a wrench for them. The old place is all they've ever known." I called them from their homework, and we held a family conference all higgledy-piggledy on our bed.

"Tonight's question," I said, "is whether you want the camp or not." I stopped their enthusiasm with, "Now wait! Don't just say yes before you think what that would mean. If we got the camp, we wouldn't read books together any more in front of our marvelous old fireplace in the evenings with cocoa and popcorn. No more breakfasts out back. It would mean leaving those trees, Richie and Mark, that you helped plant. And your rooms! You won't ever sleep in those rooms again because somebody else will be sleeping there. Buying the camp means leaving that, all that! *Everything!* And we couldn't go back, either. So think before you say anything!" After a long silence, Miranda began sniffling.

"Couldn't we . . . couldn't we . . . well, keep the old place?"

"We'd have to sell it," I answered.

"Then . . . then I don't want the camp!" she blurted in a rush.

"And the diving tower too, I guess," added Richie wistfully.

"The diving tower too," agreed Elisabeth. "Everything!"

"But I can remember painting that diving tower!" Richie objected.

"And Dad built that whole great dock," Elisabeth exaggerated sentimentally. The tears began rising to my own eyes. "All that would belong to somebody else."

"Would *not!*" argued Mark. "When I pull up there with my grandchildren, I'm going to tell them that it was me hammered the nails in the topping."

"But the boathouse!" urged Richie, "and those cliffs! Where is there anything? . . . *anything* like them? I'm for going."

"Me too!" chorused Mark. "Varoom—varoom! Me and my sixty, boy."

"Me too," agreed James quietly.

"Bet you'll stop at the old place with your cruiser, James," said Richie.

"I will," he said.

"But maybe they won't let you," said Mark. "Then what would you do?"

"I would anyway!" he said simply.

"Miranda?" asked Elisabeth.

"Don't want to go," she answered, still sniffling.

"But the camp, Miranda," Richie argued, "the camp is so much more beautiful . . . and such an *investment!* Where could you top it?"

"What's an investment?" asked Miranda suspiciously. Then, "Well, I don't care what it is. I don't want to go anyway, so there!"

"You mustn't get too upset, love," said Elisabeth, giving Miranda a gentle pat, "we haven't got the money anyway."

"Yeah," put in Richie, "where *are* we going to get the money?"

"Don't ask embarrassing questions, dear," said his mother. "It's bad manners."

I didn't have that kind of money, but I did scratch up a dollar and a half to call Mr. Birchard, remind him of my interest, and raise my offer to thirty. He received that incredible figure with what seemed to me incredible indifference. He said he would discuss it with the other camp directors. He did, and they rejected my offer. Then it was my turn to say, "Now *that* is really and actually *that!*"

"I'm glad it's over," Elisabeth sighed. "Where could we ever have found that sort of money? Besides, I sort of feel like Miranda. It was a nice dream, though. Anyway, it will be a lot easier for you, won't it, darling? You know the old place backwards."

"That's right," I agreed.

"And that's good, isn't it?"

"Sure," I said sadly. "Sure it's good!" I remembered the sense of space between the soles of my feet and the dock on which I stood. "But the next question is whether we can go back."

"We never left," she objected.

"Maybe not," I pondered, "maybe not. We'll have to see."

On a bleak winter afternoon, I sat in my office idly fiddling with papers and wondering how to abolish February when the telephone rang. I picked it up. It was Mr. Birchard in Kingston—Kingston, Ontario! He was hemming and hawing about my offer, hinting at some difficulties they had experienced with their other prospective buyer, and was I still interested. As I automatically said "yes," (I can't get over the habit of saying yes on a telephone) a thirty-thousand-dollar ax suddenly reappeared, poised over my head. Having disposed of it months ago, I had forgotten its terrors.

"Well," Mr. Birchard went on, "if you could see your way clear to up that to thirty-five, I feel sure we could do business." I laughed.

"Mr. Birchard," I confessed, "I can't see my way clear to the thirty, let alone thirty-five . . . not at this very moment, I can't. I'll have to think about things and call you back." Hanging up, I left the office; I suppose I must have walked home, but I don't remember the trip.

"You're kidding!" Elisabeth exploded. "You must be kidding!" Then, after a pause, "Anyway, we can't, and that's an end of it. I know you're going to say it's only money, but we can't."

"I'll go to the bank tomorrow and find out if we can't," I answered.

Next morning I sat in the bank before Mr. Burns's desk, where I explained the situation, calm in the assurance of his laughter when I had finished. Then he said with equal calmness, "I don't see any problem here, Mr. Russell. When would you like the money?" My heart sank. Betrayed, deserted, helpless, alone. Bank policy! I had counted on that bastion of conservatism, of common fiscal sense, and it had given way at my first feeble thrust.

"You mean," I quavered, "you mean you'll actually lend us the money?"

"Of course you'll have to make a formal application with Mrs. Russell's signature, and it will have to be approved by the loan committee, but I don't foresee any problem there."

"Well, I do. I see lots of problems."

He chuckled good-humoredly.

"Oh, we're not going to worry about you, Mr. Russell. You're a pretty good risk."

"How do you know a thing like that?" I pressed. "How can you?"

"We have your record, you see. We know you pay your bills."

"That's not me," I urged. "That's her . . . it's my wife. She's always paying bills. She's obsessive about it. That's not *me!*"

"Well, there you are," he laughed.

"No. There *she* is," I countered.

"But, we shall want her signature too," he said kindly, as if speaking to a small child; "so everything will be all right. Here's the form. Take it home and fill it out and let us know when you would like the money transferred to your account . . ."

"But I don't really understand," I said almost desperately, "why in all common sense would *you* lend *me* this kind of money?"

"That's what we're in business for, Mr. Russell," he explained gently. "You see, we lend money in order to get the interest. So if you'll just fill out the form, everything will work out fine. Good day, and thank you for coming in."

Defeated, I returned with the news.

"You mean there's no way out?" she declared.

"We don't actually have to buy the bloody thing. Not unless we want it, that is."

"Oh, God! Not *that* again! But how will we ever pay it off? That's what I want to know. How?"

"With you in charge of the checkbook," I said, "anything's possible, and he was very firm about having you sign. Besides, we're both teaching." This was her second year of full-time teaching at a local private school, and she was chairman of their English Department. "What we do, you see," I explained, gathering strength, "is, we borrow expensive money and pay back with cheap money, because our salaries are bound to go up over the years. Sound economics, eh?"

"Economists! What do they know? Let's fill out the form, though. Maybe the committee will turn us down." We did, and they didn't.

"Good," said Mr. Birchard calmly. "I'm sure the other directors will agree."

I suggested July first as the settlement date, and he said good again and that he was sure the other directors would agree. In fact, to everything I said, he said the same thing.

"He was awfully agreeable," I said to Elisabeth.

"He's getting the money," was her reply.

"Varoom!" cried Mark.

"Hurray!" shouted Richie.

"Good," said James. Miranda said nothing.

"You'll get to like it, love," said Elisabeth gently.

"You can maybe even have your own cabin," Richie helped. "Think of that!"

"Don't want a cabin," she sniffled. "I like my old room. I like the old place!"

"We all like the old place," I said.

"Then why are we leaving it?" she asked.

"Because the new one will be nicer," I said.

"And how do you know that?" she pressed.

"I don't," I confessed a little sadly. "I really don't. But I hope it will be."

"Gonna be great!" cheered Mark.

"Varoom—varoom!"

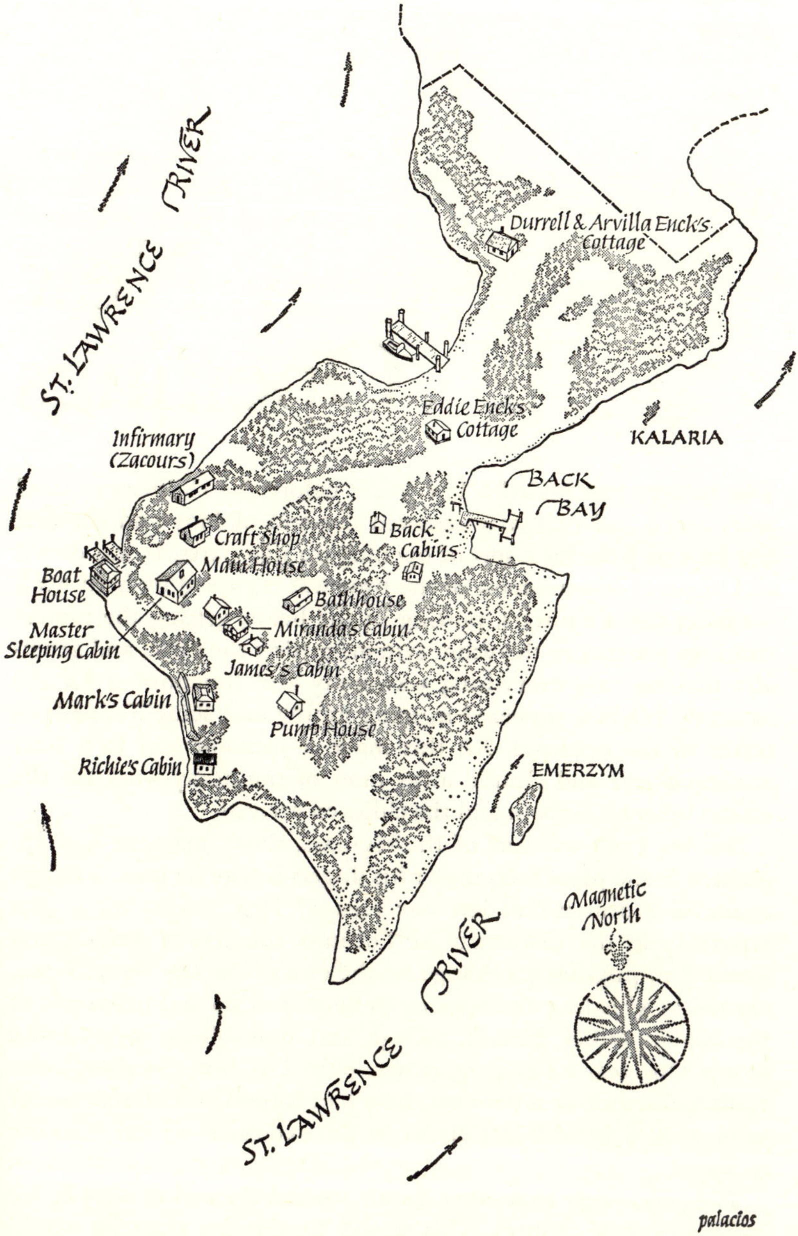
ST. LAWRENCE RIVER
Durrell & Arvilla Enck's Cottage
Eddie Enck's Cottage
KALARIA
BACK BAY
Infirmary (Zacours)
Craft Shop
Back Cabins
Main House
Boat House
Bathhouse
Master Sleeping Cabin
Miranda's Cabin
James's Cabin
Mark's Cabin
Pump House
Richie's Cabin
EMERZYM
Magnetic North
ST. LAWRENCE RIVER
palacios

12

New Openings

The college year ended a few weeks before the children's school year, and I had regularly reserved the end of May and the first few days of June for a sort of opening-up stag party. It was only a sort of party because most of the time was occupied by hard work —taking down shutters, putting in the boats, getting the motors running, putting in the water system, cutting the grass, and generally making sure everything was ready and functioning for the summer. We also squeezed in some fishing and some five-and-ten poker in the evenings. All this made a pleasant end to a busy academic year and cleared away most of the heavy work for the season before Elisabeth and the children arrived.

Durrell Enck was one of the regulars. Every machine and appliance in the place sang under his fingers before he quit. Though medieval Norman had left our History Department for a professorship at the University of Toronto, the end of May always found him jingling his change on the end of the Bay Street Dock, nervously whistling and waiting to be picked up and taken out to the island. Other friends, college and non-college, swelled the group. One year a couple of miners joined us, brought along their equipment, and, in a few days, they had drilled and blasted out of solid rock a twenty-foot basin as an extension of my docking facilities.

The good cheer that filled the air toward the end of May in '68 had its edge of sadness. This would be the last time we would open up the old place. I said something about it to Durrell as we pulled into the parking lot at the Bay Street Dock in his heavily laden station wagon.

"What's the difference?" he asked callously. "Next year we'll open Mohawk, that's all." But for me there would be a difference.

As I stepped from the water taxi onto my dock, I felt a pang of guilt beneath the mingled joy and sadness of my last return. Down the catwalk under the willow tree to the shore and up the worn path through the already tall grass to the front steps, I came to the memory-filled house that had waited my coming through all those months of storm and winter silences. I came this time as its betrayer, and, since it contained so much of me, my own.

We took off the pair of shutters from the double doors in front, opened them both, and dragged the fiberglass from the living room out onto the porch. As I stood halfway down the steps holding the bow to help ease it down onto the grass, I stroked its smooth rail and said: "At least you're coming along to the new place."

"If it's anything I can't stand," growled Durrell, straightening up at his post by the stern, "it's a goddamn *sentimental* millionaire!"

"How about if he's in debt two million?"

"They're even worse! Come on. Here she comes," and he pushed the boat forward until she tilted on the top step and I took her weight, going slowly down onto the grass.

That was Thursday. By Saturday noon things were well along, and four or five of us were having a sandwich when the phone rang. Elisabeth's voice, strained and shaky, said: "Trouble about Richard drinking. You'd better come." Her voice serious, she explained. "He's going to be all right," she concluded, "but can you come?" Carl was driving back on Sunday.

"I'll be back tomorrow," I said. "Try to take things easy."

"Everything all right?" asked Durrell when I had hung up. He knew it wasn't. I told him as the others listened.

"If you want to leave now," Durrell offered, "I'll run you down." Casual, the "running me down" those four hundred miles.

"Tomorrow will be all right," I answered.

Next day I hummed down to Lancaster with Carl.

"He's all right," said Elisabeth at the door, "but this is only the beginning. Mr. Bristol was here a little while ago." Mr. Bristol was the headmaster of the school at which she taught and that the children attended. "He feels the school is involved so much that he doesn't think Richie ought to go back there next year."

"What? What's he talking about?" I exploded.

"I told him," she explained calmly, "that if Richard couldn't go back, neither would the rest of us, including me."

"And what did he say to that?"

"He said he would give us his final answer tomorrow, but, darling, I'm not going to teach at a school that's too good for my own son. Crazy or not, that's how I feel."

"Then that's settled," I said. "It's up to Bristol."

"And jobs aren't easy to come by," she mourned. "You realize that, don't you? And there's Mohawk. It's not too late to stop the whole business . . . I mean, we haven't actually borrowed the money yet, so . . ."

"We *could* cancel it, but we won't. We'll be in worse shape than we had expected, but so what? We'll manage. The school's going to suffer more than we are." It was no secret that she was an outstanding teacher. The students knew it too. They respected her as much as they adored her.

Next day Mr. Bristol appeared and explained that, from his point of view, it would be best if Richard went elsewhere in the fall, and he would do all he could to get him into the best school in the country. We thanked him for his good intentions.

Elisabeth gave him her resignation and we withdrew our children.

Bad news spreads instantly, and so did this. Our telephone and doorbell jangled for days with students, parents, and other teachers trying to talk Elisabeth out of her decision, but her answer was the same to all: "Mr. Bristol is doing what he thinks is necessary from his point of view, and so must I from mine."

As a tribute that only another teacher could fully appreciate, the entire senior class left their graduation dance and packed our sitting room to express their appreciation and admiration for her teaching. We almost felt sorry for Mr. Bristol in his presiding for an hour over his empty gymnasium.

"But what I don't understand," Richie mourned later, "what I don't see is why Mark and James and Miranda have to leave, and why does Mom have to go and quit her job? It was *me!* It wasn't *them!* Why should they . . . why should everybody have to take it in the neck?"

"Because we're a family," I tried to explain. "Whatever happens to one of us—good or bad—affects all of us. That's what being a family means."

"But how are we going to pay for the camp now?" he pressed.

"Same way as before," I said. "Just take a little longer, that's all."

"But wouldn't it be easier if we forgot about the camp?"

"Sure. But then we wouldn't have it. Now, here's my moral, so listen! If you want something, you don't quit just because it's hard to get. Sometimes we've got to quit, but it's a bad habit to get into, and we're not going to start now. That's moralizing, I know; but I happen to believe it."

Would what I had said be any use, any help to him? Would such words from my parents have been any help to me? Would they have blunted those spears of secret anguish on whose points I had squirmed during my teens? I doubt it. Perhaps they had doubted it too, for they kept their silence, putting their faith in the instructive power of action—what they did; how they lived. Were they right? Were they wrong? I don't pretend to know. Wise or foolish, I had said to my own son what I could, what I had to say, and we did what we felt we must do.

By mid-June we had cleared what could be cleared, and set off for the river. Durrell and his friends had put the finishing touches on the opening-up. Everything worked.

"The perfect time to sell," I cheered.

"But suppose nobody . . ."

"They will," I said, "and what's more, they'll pay cash!"

The directors of the camp had told Whiz we could have the keys even though settlement was a week off, and we had spent a day inspecting our future.

"The only trouble is," Elisabeth complained that evening after our return, "with all those buildings . . . how many are there, anyway?"

"I have no idea," I replied.

"Well, however many there are, there doesn't seem to be . . . at least I can't remember a single, as you might say, *house*—a place to live, if you know what I mean."

"I don't," I admitted. Though we had spent the day tramping

from building to building, the layout, so to speak, was one big blur in my mind.

"Well," she said with a sigh, "maybe if I try to explain it again to you . . . maybe I'll get a better picture myself. So. Let's start with the boathouse. You know the boathouse, don't you?"

"Yes." But that wasn't true. More precisely, I was afraid I might. I could remember stepping inside it that first time and how bouncy the floor felt. Whiz had noticed and said it was nothing to worry about . . . just a timber gone. All a fellow'd have to do was shove another one under there, and I hadn't had the sense to ask him how, exactly, one "shoved another one under there" when it was a two-story affair maybe fifty feet long and thirty feet wide. But anyway, I said yes, I knew the boathouse.

"Well, you go up from the boathouse along this little path till you come to . . . to the kitchen. It's not just the kitchen! We're going to have to give these places names so we can understand each other. Let's call that kitchen affair the Main House, even if it isn't a house."

"Seemed like a house to me," I objected.

"It isn't. No bedrooms; no bathroom. There's just that huge kitchen, then the monstrous screened-in porch they used for a dining room; then the little indoor dining room and the big pantry with that awful walk-in refrigerator. We'll have to tear that out sometime. So that's the Main House. Then you go . . ."

"Why not just put some beds up there?" I asked. "What's a bedroom anyway? . . . it's only a room with beds in it." But that didn't suit her.

"No, no! We can't do that. We'd be all scrunched up, and God knows, there're plenty of bedrooms. You just go along a little ways to this big cabin that has three rooms and a big bathroom."

"Go first class," chimed in Richie. "With the names, I mean. Call that the Master Cabin."

"Okay," she agreed. "And then there is this sort of string of four little cabins after that. . . ."

"One for each of us," said Mark.

"I got mine picked out," said Rich. "Wrote my name on the door today. It's the last one way down on the point."

"And mine's the one just this side of Rich's," said Mark. "You can have the next one, James, and Miranda, yours can be the one next to Mom and Dad."

"Don't want a cabin," said Miranda.

"So then," continued Elisabeth, "sort of behind Richie's cabin, there's the pumphouse with all that junk . . . the tank and so on. Well, all that whole string is off to the right, if you're facing the shore from the boathouse. Then, if you come back and go off to the left, the first building back a ways is what they called the Craft Shop, where the girls made their belts. To the left of that is a funny sort of motelish building, the Infirmary, with a line of rooms all opening onto that long porch. . . ."

"That whole porch is rotting out too, Dad," said Richie cheerfully.

"Be quiet, Richard," said Elisabeth, "I'm trying to explain things to your father. Then, on past the Infirmary, a long way past, and down the hill in that bay is the Caretaker's house."

"If that's a house," I said, "we'll live there."

"It's too small," she said. "Besides, it would do nicely for the Encks, for Durrell and Arvilla, and they could put Eddie in one of those cabins out back."

"Out back?"

"Well, yes. On the other side of the island, there're three or four of these newish cabins . . . all rottenly built . . . probably put up by the camp. Then, scattered all over the show, are these big tent platforms. How many buildings does that make?"

"Haven't the foggiest," I replied.

"But you were supposed to be *counting!*" she exasperated.

"Impossible!" I said.

"And you forgot the big bathhouse, Mom," put in Mark. "You know! that place with all those sinks and toilets?"

"The point is," she said, ignoring the interruption, "the point is that, in all that . . . in all those places, there's nowhere to live." The fact seemed to have her in its grip by the end of her explanation, and she turned to the children with her mournful appeal. "Your father, children . . . your father pays thirty-five thousand dollars! Get that? Thirty-five thousand dollars we don't have. . . ."

"We'll have it next week," I interrupted, but she went right along.

"And for what? For the privilege of moving out of a perfectly comfortable friendly old house to a bloody great estate with twenty buildings, not one of which . . . not one of which can be lived in! Not *one!*" I broke the silence that acknowledged the force of her rhetoric.

"Nonsense! We've got umpteen cabins to sleep in."

"Sleep!" she fumed. "You can't even move in those cabins, if you'll just remember, because they're all chock full of those damned double-decker bunks!"

"So we'll sleep in them," I pointed out.

"We can't!" she said. "Those places all have to be cleared out and given a good scrub."

"Then we'll start in the morning," I said.

"Wait!" cried Richie, "quit shoving. It won't go through the door. It's too wide." I put down my end of the ungainly iron skeleton.

"It's got to," I said firmly, bumping and scraping my way along and between the scaffolding of other bunks toward the door.

"It's no good!" he repeated emphatically. "It really isn't, so don't bother trying to get over here. Even if you could, it's not going to go through. Only way is to take it apart, that's all."

"It!" I yelled, "it and about a hundred more. It's got to go through," and I picked up my end and drove forward.

"Cut it out," he howled. "You're squashing me. It's just no good! Can't you get that through your head? We've got to take them apart anyway to store them. We can't just shove them outside, and there's no place big enough to house them if they're all assembled . . . no place, that is, except right where they all are right now." He was dead right, but that didn't make the fact any easier to swallow.

"Screwdrivers and wrenches, then," I said, sinking down sadly on a sharp bedrail.

"Cheer up," he threw back as he left to get the tools. "There're actually only about fifty." Statistically his figure was probably closer, but my hundred was spiritually more accurate.

Before that melancholy morning, I had frequently been struck by the virtues of metal construction. It had seemed so smooth, so secure, so permanent. I came from that encounter with those fifty sets of double-decker bunks a changed person, a strong believer in the reasonableness of wood. Wood is something that responds to a saw, an ax, or a sledge hammer. It consents to destruction.

Metal, on the other hand, is sullen, defiant. Discarded though it might be in a clump of bushes—and there were plenty of inviting clumps around—it would lie there for years, maybe eternity, to grin its rusty grin of disfiguring triumph at all who passed. I had known fences, barbed-wire fences, like that. Their useful life long since over, they would first sag for years between rotting posts, then subside into the tall grass where they tenaciously preserved their teeth and mesh for the bare legs of picnickers.

No use to plead "It's not my fault" to such fences or to the silent gallery of scaffolding about me. It was not I who had use of them, who had smelted them out of the rock and dragged them here and slept in them for ten years. But the responsibility for getting rid of them was ours and they would wait, if necessary forever, until we did something about them. For that privilege we were paying handsomely—mortgaging years of our future. Along with those rocks and tree-lined shores, the sound of waters, the song of birds, and the feel of sun and wind—along with all that loveliness, we had bought this responsibility, this drudgery.

Feet thumped on the porch and the door opened.

"Here you are, Dad," said Richie, leaning through the square arch of iron to hand me the tools and bumping his head.

"Got a set for yourself?" I asked, stretching out to take them.

"No," he answered.

"Then bloody well go and get some. Pollution is everybody's problem," and I began on the first rusty bolt.

I used to admire workmen who scorned nails, even screws, and went in for nuts and bolts. I rejoiced in the stability and permanence of their work. Silly child! I soon found that stability and permanence had other edges. Nuts and bolts, however securely they may bind at the outset, have a way of gently loosening till the members they secured wobble and flap. Then these once-noble nuts and bolts rust, rust so firmly that they achieve the ideal marriage—what once was two become perfectly one. So bonded, they can repel the most secret, divorcing influences of the most expensive penetrating oil. They can be put asunder only by the merciless blade of the hacksaw. While they shed no blood of their own in this process, they exact it from the knuckles and fingers that drive the blade back and forth between the iron struts they were meant to hold firm.

And this, I reflected gloomily, was the sort of job I was supposed to be good at—one requiring almost no intelligence or imagination, no ability to analyze or grasp intricacies of mechanical organization—one demanding merely a sort of Chinese diligence, the kind that, handful by handful, had built the Great Wall. Determination, Durrell was fond of saying, takes the place of a lot of talent. Those successes of mine not attributable to accident had come, or, perhaps more accurately, given in from the sheer weariness under a mindlessly persistent attack. These fifty double-deckers put me to the severest test, success being the hollow victory of an empty room.

After two days of carting dismantled bunks to a back cabin over a couple of hundred yards of rough ground, the children gave out, complaining unreasonably of stubbed toes, barked shins, bruised knees, sore hands, aching backs, and so on and so on. They straggled off, muttering rebelliously about how this was supposed to be a vacation, wasn't it, and they were going swimming. Elisabeth and I were left stumbling along over the rocks and tree roots at either end of our rack or stack or pile of jagged, heavy, sharp, unforgiving iron. Tripping on a log, I went down, and brought her with me in a patch of wild raspberry bushes. Too weary to swear, we simply sat.

At last she broke the silence. "Why . . . oh why do we do all this?"

"Same reason," I answered weakly.

"Same as what?"

"Same as before. Remember? We're the sort who can't help it."

"Other people can. Why can't we? Other people live decent, quiet, reasonable sorts of lives. They don't go bumbling about in bushes carrying stacks of great heavy iron beds to jab them and poke them and hurt them when they fall. Other people have TV. They watch it. Some go on cruises. They take drives in the country and go to the theater. . . . I love the theater. How come we don't do those things? We could do those things, but here we are sitting in the brambles with these . . . these *things* digging into us, thorns and sharp hard corners digging into us all over. We supposed to be enjoying ourselves or something?"

"I suppose we must be," I answered. "Cheer up, though. Can't all be like this. Anyway, we have made progress. Haven't we?"

"The theater . . ." she went on dreamily, ignoring me and my progress. "And we were actually at Oxford . . . that beautiful, beautiful city . . . libraries, concerts, plays. . . . All glorious, it was, and then I had to go and marry *you!* You said . . . you said you had a future to offer"—gathering steam—"and here we are, fifteen or is it thirty—seems like a hundred—years later here we are sitting in a brier patch! a *brier patch!* Oh, God."

"How about a divorce?" I suggested.

"We couldn't even get that," she wailed, "without climbing out of these raspberries! And then, even then we couldn't just leave these heaps of junk here. We'd have to put them away. So we might as well get on."

"We're almost there," I encouraged, struggling to my feet. "And I bet . . . I bet we'll be all finished with these bunks tomorrow!"

"This is no time to be cheerful!" she snapped. "Let's go."

That night, instead of counting sheep, I tried dollar bills to see if I could get a clear idea of just how much thirty-five thousand came to. Once past a hundred, the numbers became meaningless, so I tried thinking of it in terms of automobiles. Thirty-five thousand was the same as maybe twenty VW's like our bug parked over at the town dock. Twenty. Twenty cars. The price of twenty cars. Surely that must be enough to go into business with! And businessmen got rich. What's more, none of them I had ever heard of had to take apart fifty sets of double-decker bunks and lug the obstinate, angular remnants through the woods to a back cabin. Why hadn't we . . . why couldn't we go into business? Andrew Carnegie got rich, and he didn't start with the price of any twenty VW's. He probably had a paper route or something. . . . But then he didn't have any bunks . . . or sheep. . . . Steel, but no beds . . . no raspberries or divorce . . . no talent . . . just determination . . . no vacation. . . .

Next morning, at breakfast, at the old place we were still using as our base of operations, everyone seemed a bit stiff and glum.

"Know what?" I threw out, "we could go into business." No one seemed interested. "And make money," I added. Still no interest. "Get rich!" I urged.

"Yeah?" inquired Richie absently. "Could you pass the jam, James?"

"Look," I squeezed through his chink of interest, "we've got about a mile of shoreline. You know what shoreline's going for nowadays? Per foot? Per twelve lousy inches?"

"How much?" he asked carelessly, and he lathered the jam on his toast.

"An ad in the paper last week. You can check it. Two . . . hundred . . . dollars!" He dropped his knife.

"You mean per *foot?*"

"Per *foot.*" I said. "Think about that for a minute. We've got . . . well, we will have . . . about a mile of shoreline with Mohawk. You just drag out your slide rule and multiply two hundred by five thousand. What do you get?"

"My God!" Richie exploded. "That's a million . . . a million dollars!"

"You mean we're millionaires?" cried Mark in disbelief.

"It's just your father talking," said Elisabeth.

"We could be," I continued. "All we'd have to do is break Mohawk up into lots and sell them off."

"Sell them off," Elisabeth put in, "and get that two-hundred dollars a foot."

"Suppose we only got a hundred," Richie rejoined. "That's *half* a million . . . five-hundred thousand. Not bad, really. Invest that at five percent, and you get twenty-five thousand a year. I could use twenty-five thousand a year."

"To cut it up, though," objected his mother, "you'd have to get it surveyed, and that, I can tell you, would be expensive!"

Richie collapsed with laughter. "Mother," he patronized when he had recovered, "you're so small-time! You got to learn to think big, like Dad! *Expensive?* Cost maybe two or three thousand to have the place surveyed, and what's two or three thousand?"

"Then we could get a cruiser," said James with a trace of eagerness. "Go off for weekends."

"And not tell your wife, eh, James?" laughed Richie.

"Well, anyway, we could," James replied staunchly.

"And we could stay right here, Miranda," I went on. "Not move at all . . . not sell this place."

"Good," she answered. "Why don't we do that then?"

"Stay here and get our own sixty," Mark chimed in. "Varoom!"

"Why not?" I asked.

"You're not serious, darling, are you?" Elisabeth asked nervously.

"That's what people do," I said. "That's how they get rich!"

"Some people," she said contemptuously. "De*velopers!*"

"That's right. Developers. They buy big chunks of land, break them up, and sell the lots."

"But not *us,*" she said.

"Why not?" pressed Richie. "We could found a family fortune . . . the Russell Dynasty!"

"Be no sweat," I said. "The deal's not even finished, and two or three people have already approached me for lots."

"Then sell!" directed Rich.

"We'll do no such thing!" his mother declared.

"But why ever not? Like Dad says, we could be *rich!*"

"All right," I said, "we could, but what, exactly, would we do with the money?"

"Same thing everybody does with money," he returned. "Spend it! Get things they want."

"And what do you want?" I asked.

"Swimming pool," Richie said.

"You've got the whole St. Lawrence down there at the end of the dock," I pointed out.

"Then how about a car?" he suggested.

"There's one over at the Bay Street Dock."

"A sixty," Mark declared enthusiastically.

"That's one thing, but after that?"

"A cruiser," James offered. Everybody laughed.

"How about a place on the St. Lawrence?" Richie threw in. More laughter, then silence. "But," Richie added in some distress, "there's got to be *something* to do with a million dollars. Everybody seems to want it bad enough!"

"Bad enough to do what?" I inquired.

"What do you mean?" he said.

"Well, bad enough to do what to get it? The question is simply, what's money worth. What is a person willing to do or give up to get it."

"If we didn't have any, it would be worth a whole lot," Richie said sharply.

"That's true," said Elisabeth, "if we didn't have any, but the point is that we do have some . . . quite a lot, really. We've got a house, a car, and food. That's a good lot. Plenty of people have less. But the question is what would we do with more? Would it be worth breaking up Mohawk to get it? Exactly how would we be better off if we did that?"

"Impress people," suggested James after a pause.

"But we don't care about that," Elisabeth replied.

"Oh," exclaimed Richie with annoyance, "you two are just too virtuous for words."

"It's got nothing to do with virtue," I objected. "It's just the question of how and where we want to live, that's all. We're buying Mohawk because we want to live there. Nobody seems able to believe that. People have been asking ever since the news got out that we were buying it. Everybody wants to know what we're going to do with it. When I tell them we're just going to live there, they don't seem able to understand the language. They don't actually call me a liar, but they obviously don't believe me."

"They can't believe you're dumb enough not to make money out of it," said Richie.

"But that's exactly what we are!" declared Elisabeth. "We're that dumb!"

"Everybody has to decide the same question for themselves," I said, "how they're going to live, what they value, and what they're going to do or give up to get it. You kids are going to have to decide that too."

"But we could hire people to build docks if we were rich," Mark suggested.

"I like building docks," I replied.

"And carry bunk beds," he added.

"I don't like carrying bunk beds," I admitted, "but it's got to be done; so, come on. Let's do up the dishes and get back to work."

"Trouble with us," observed Richie, "is we seem to be rich already."

"We are," I agreed. "Hurry up. We've got lots to do."

As each cabin was cleared, Elisabeth attacked it with broom and mop and buckets of water dragged up from the river and heated on the huge bottle-gas stove in the kitchen. As each was finished, parties of scouts were dispatched to scour the premises for chairs, chests, and single beds for furniture. These discovered, the wheelbarrow was brought into play, and, heavily laden, it groaned and thumped along over the rocks and roots before its groaning and thumping wheeler. After two or three cabins were fitted out for Russell residents, we heard the cheerful clanging of the camp's old ship's bell. Instead of announcing to a hundred campers that it was time to switch now from nature study to leatherwork, it called our band of weary workers to the little indoor dining room for tea—our first meal at Mohawk.

We crowded around the table in the dining room off the huge kitchen in the Main House. The tea, as always, was strong and hot; the fruitcake, as always, was its reliable, stodgy self with its occasional raisin surprise. With the familiarity of our ritual, we combated the unfamiliarity of the setting.

No familiar sun, no wind, no friendly creak of front-porch steps beneath us, no catbird's song from the jungle of scrub off to the left. Around us four strange walls, strange chairs, a strange table, strange cups. How long would it be until we entered this future that surrounded us? Can the future only be entered when it is transformed into the past? Our scratches, bruises, and aching muscles were proof of our efforts to earn a sense of ownership over this newness, but time had not yet set his comfortable seal of familiarity upon these surroundings; so we seemed only visitors. Like visitors, we were ready, even eager, to admire everything, but also, like visitors, a trifle uneasy, a little anxious to go back home. In the pauses of conversation, we all seemed to be listening for the footsteps of our approaching host, to when we could make our apologies for having intruded before withdrawing. But they never came. How long would it take, I wondered, before our footsteps would become those for which we waited.

"Hey!" cried Richie, getting up and going to a sort of corner alcove, and he read aloud; " 'Mary Lewis, October 2, 1920. Tom Lewis, October 2, 1920.' And there are marks here on the wall to show how tall they were right up to 1944 for Tom. And here's Frederick Lewis Jr. And listen to this; everybody listening? Here's E. Paul du Pont III, August 31, 1946 with some more for E. Paul; and here's A. Felix du Pont IV, August 31, 1946, all the way up to 1949. How about that? Du Ponts! The Russells have finally arrived!" and he blew an imaginary fanfare. "We own one of the Du Pont estates, no less!"

"Not really," I dampened. "It was built by the Lewises around

1917. There was a story, though, that they were related to the Du Ponts. Sounds like she must have been. Those kids were probably her nephews. Like I said, we're already rich."

"Bet those Du Pont kids didn't have to carry any bunk beds," said James testily.

"Maybe not," I admitted, "but you were unlucky enough to have a pair of Oxford washouts for parents; so finish your tea, and we'll get back to work."

"I bet this used to be the servants' dining room in those days," mused Elisabeth.

"Things haven't changed much," Mark observed, "except we pay the bills too."

"Speaking of work," put in Elisabeth, "cleaning up would be a lot easier if you chaps could figure out the water system for this joint. Lugging buckets from the river isn't . . ."

"Yeah!" said James with genuine warmth.

"You put in the water at the old place," she urged, "so this shouldn't be hard."

"I understood that one," I countered, "and even then, it was never simple." Still, after tea I collected the pipe wrenches and took Richie and Mark, or rather they took me, down to the pumphouse more than a hundred yards away. Inside was a thousand-gallon pressure tank—very empty—and beside it, squatting like a deformed bulldog, was the pump.

"This," I said authoritatively, rapping the tank, "is the tank. And this," bending down to give the pump a friendly pat, "is what we call the pump!"

"Hurray!" they cheered. "Now what?"

"Well . . . what we have to do now is . . . we have to trace the intake line down to the river," and we walked along a one-hundred-fifty-foot length of inch-and-a-half galvanized pipe to its end, which jutted out over a small cliff at the water's edge.

"Got to be another piece around here that they shoved out into the river and attached to this," I said.

"Maybe this is it," suggested Mark, tapping another piece of pipe. "It's got a foot valve on the end, but it's about forty feet long. We'll *never* be able to get that going out at the right angle so you can make the connection between them watertight!" Now, never is a long time, but it seemed almost that long before I could mesh the threads of the union connecting the two. The rocks on which we worked were covered with wet moss, and the wake of every tour boat and cruiser soaked us to the knees.

"That's enough for one day," hoped Richie as I gave the last twist to the union with the big wrench.

"We've got to prime the pump and get it working," I corrected.

"What?" exploded Mark. "You don't know where to prime that pump, Dad!"

"Nope. But we can find out," I said. "Come on."

"Be reasonable for a change," Richie complained. "Just be reasonable. Suppose we did get that pump working—we won't, but suppose we did—where would the water go? *After* the tank, I mean?"

"To the kitchen," I said.

"Dad!" Richie howled. "You don't understand! There are pipes, millions of them, going off to every one of these cabins, and they go way past the kitchen up the hill to the Infirmary and over the hill all the way down to the Caretaker's."

"That's the whole point!" I explained. "We have water everywhere."

"Everywhere is right," he laughed, "but not in the cabins. This system has all been drained so the pipes wouldn't freeze. There must be seventy-eight low spots, and there must have been a drain plug at each of them. We don't know where Whiz's box of plugs is, and if we did, we wouldn't know where to put each one of them, and you've got to put them in before the system can work. We've got to get Whiz to help on this one. He knows the system. We'll go along with him and learn it, make a map of it so we won't need him next time, but we can't do it without him this first time. Those pipes snake around this way and that . . . some go underground . . . and well, it's impossible, that's all. We could never figure it out ourselves."

Sometimes even a professor has to give way in the face of truths, and my son was speaking it. Sadly I picked up the wrenches and we trudged back to the Main House.

"Well done, Rich," congratulated Mark, and they disappeared toward the boathouse for a swim.

"Any luck?" inquired Elisabeth, stopping her mopping. "I turned on the taps, but nothing's happened."

"We need Whiz," I said, putting down the wrenches and sinking into a chair.

"Whiz? But you used to put our water in," she argued.

"That's different," I answered.

"How did you get so wet?" she asked.

"Putting in the intake pipe."

"Then it's all set, isn't it? Why can't we have water?"

"It's not that simple," and I tried to explain.

"But I thought you knew about pipes and things," she said, returning to her mopping.

"No."

"They hard or something?"

"Yes."

"Then how about getting me another bucket of water. I'm almost finished here. Shall I leave the taps on just in case . . . in case something should happen? We won't have a flood or anything, will we?"

"No danger."

"Then if you'll get the water, we can finish up here and go home."

I called Whiz that night, and he agreed to come the following Saturday. Fortunately he came early, because it took us the whole day to get the affair operating. That evening water did indeed come pouring out of Elisabeth's taps. Richie had been resoundingly right about the complexity of the system. Plugs had to be inserted in pipes secreted under cabin steps and in just plain holes in the ground, which were concealed by innocent-looking boards. But even after the water was operational, Richie complained that the toilet in his cabin leaked.

"Tough," commented Mark, who had no toilet in his. Richard's had been occupied by the woman in charge of the camp, and that explained his luxuries—a toilet, bathtub, sink, and hot-water heater. The other cabins had been well fitted out with plumbing under the Lewises' regime, but the camp had dismantled what they could in order to get in more double-decker bunks.

"All right," I said, picking up the wrenches. "Let's have a look," and we went down to the rocky point where Richie's cabin stood. Kneeling before the bowl, I soon found the leak and a couple of twists with a wrench miraculously put a stop to it, but, as I worked, I knocked over a box that had stood behind the toilet. When I had finished, I pulled out the box.

"Holy smokes!" shouted Rich, "that's Ballantine's Scotch." I shook it, and it emitted a promising gurgle. Opening the top flap, I reached in, and there, indeed, was a bottle, a forty-ounce Canadian quart of Ballantine's. My surprise and delight suddenly faded as I wondered why she had hidden it there . . . for what special occasion had she bought it to celebrate? No ordinary one, surely, or she would not have hidden it. And that occasion, that event, that meeting . . . it could never have taken place, for the seal on the bottle was still intact. What sadness must she have known?

All of us on the island had met her. She had carefully made the rounds, introducing herself to us all as a matter of good public relations. Most politely, she expressed the hope that the campers would cause us no inconvenience. Pleased though I was to have found the Scotch, I was embarrassed, too, for I held, as it were, a crack in her public life, a crack through which I could not see clearly, but a crack nonetheless.

"You can have a party with that," encouraged Rich.

"No. Not with this," I answered.

"You can't very well give it back to her," he retorted.

"Isn't that funny," mused Elisabeth when I showed her the bottle. "The other day I found some love poetry mixed up with a pile of bills. I wonder what it was all about." For others, this had been their world where they had loved and hoped, dreamed and grieved, and their ghosts moved about it still, almost more real than our bodies. Slowly those ghosts must fade as we made their world into ours. That had its sadness too, because, for them, that past was as real as our present. For what future tenants would we leave the marks of our stay?

"We won't use her bottle for the housewarming," said Elisabeth, "but we're going to have to lay in a good supply."

"Don't you want to wait till we're finished before we throw a party?" I asked.

"We'll never be finished," she replied, "so the sooner the better. We won't invite anybody, though, till we've actually bought it, but that's only a day or two."

The day or two came and went, and on July 1, 1968, the paper deed was done.

"Meet the owners!" I called to the children as we pulled into the dock by our boathouse. The children, who had been left to get on with innumerable jobs, were playing tag in their bathing suits.

"I suppose we ought to take the day off," Elisabeth conceded grudgingly, "but what are we going to serve the punch in?"

"There are three weeks to think about that. Come on. Get your suit."

"That's all very well for you to say, but these things have to be arranged! They don't just happen, you know! And this dock! Lots of the topping is rotten. Anybody could go right through and break a leg. We can put our boats in the back bay, but that dock's not much to write home about either. Can't you do something about that? And there's our bathroom! The tub looks just awful with that big piece of dirty plywood over it. Can't you reconnect the bathtub?"

"They aren't going to want to take a bath!" I objected.

"No, but it looks so terrible, and then we've got to arrange all those tables and benches! For starts we've got to clear out the outdoor dining room and get the rug down. People are going to be nosing around, you know. People just love to nose around!"

"How many invited so far?" I interrupted.

"Forty-three . . . no, forty-six. But then there're the Eddys, the Batemans, and the Stewarts, of course. How am I supposed to know how many? But they'll all be snooping."

"There's enough room. Let them snoop!"

"But that's just it! They will, and they'll all be accusing me, not you . . . they'll be accusing *me!*"

"Accusing you of what?"

"Of being a rotten housekeeper, that's what. These Americans have high standards . . . terribly high standards! Oh, I know them, all right. Everything in its place, with clean towels and clean face cloths in the bathrooms that nobody has ever supposed to have touched. Don't think I don't know about all that, because I do."

"If they don't like our towels, they can use the pine needles," I said.

"Don't be funny," she snapped. "And there's all this old lumber just lying around. Looks awful! And all those tent platforms. . . ."

Mohawk is sixteen acres, but all except the hearts of its densest thickets got some sort of attention from the corps of four children and one husband under Elisabeth's command. During those weeks, she ruthlessly put down our occasional attempts at mutiny and restored loyalties with popcorn and toasted marshmallows around evening fires.

By three-thirty on the appointed afternoon, when the first guest-carrying boat purred into our empty harbor, Mark and James, clean and respectably clad, were at their posts on the dock as official welcomers and tier-uppers. Richard, our gourmet, was in the kitchen, where he had been since morning sampling, then adding a little of this and a little of that to the cauldrons that bubbled on the big stove from which the most marvelous smells drifted all the way to the boathouse. Miranda was in charge of showing the ladies over to the Master Cabin and taking care of their jackets. Elisabeth and I, having given the last stirs to the ten-gallon tub of punch and having warned each other about its potency, prepared calmly to meet the onslaught.

By four, sixty friends were laughing and talking, milling and snooping about Mohawk. There were Americans and Canadians, financiers and fishing guides, house painters and professors. The fresh air, good food, and strong punch welded them into a roar of good humor and warm feelings.

"Great place you've got here," called a Rochester banker over the tumult. "One of the nicest spots in all the islands."

"Thanks," I shouted back. "We think so too."

"Going to run it as a girls' camp?" he queried.

"What?"

"Are you going to run it as a girls' camp?" he repeated.

"Never," I replied.

"Why not? Isn't it all set up for that?"

"Too much work!" I shouted.

"What are you going to do with it then . . . divide it up?"

"No. We're just going to live here."

"You're going to *what?*"

"Live here," I roared. *"Live here! Relax! Vacation. . . ."*

13

To Shake the Past

In buying Mohawk, we had counted heavily not only on Elisabeth's income from teaching, but also upon selling and getting a good price for the old place. On July first, Elisabeth had no prospects for a job in the fall and the old place was distinctly unsold. Hunnie had advertised it in the Syracuse papers and Cliff Carpenter had done the same in Rochester, but there had been no bites. Dismayed, I called in Ken Keyes, a Gananoque real-estate agent through whose father we had made our original purchase thirteen years earlier. Ken dropped in one afternoon to look over the property. Elisabeth was standing on the porch railing zealously painting the overhead trim. She was as demonic about sprucing up the old place as she was about setting Mohawk to rights. This required that she divide each twenty-four hours into two full working days, which she managed to do. After his tour, Ken sank into a rocker on the front porch, put up his feet, and was silent.

"Excuse me," said Elisabeth, climbing back to her perch, "but I've got to get on, you know." He said nothing.

"Well?" I inquired nervously, "what do you think?"

"Pricewise, you mean?"

"Yes. Pricewise." He rocked thoughtfully. We held our breath. I'm not sure, but Elisabeth might even have stopped painting as the tension grew. The silence getting too much for me, I said, "I mean . . . that is . . . well, what do you think it's worth?"

"Oh," he laughed, "that's an easy one. It's worth just as much as you can get for it, and no more."

"But how much is *that?*" I blurted.

"That depends," he mused. "Depends who wants it, and how much money they've got."

"You mean how much they can lay their hands on," I corrected, recalling my own recent experience.

"Same thing," he observed. They seemed quite different to me, but it wasn't worth arguing about. "Same as far as the seller's concerned. The question is . . . the question is how much to ask. That's the question, isn't it?"

"Right," I encouraged. "That's the question."

"Well," he rocked some more. Elisabeth shifted her feet on the railing, and I hoped she was hanging onto a post.

"There's two-hundred-forty feet of frontage, you know," I reminded him.

"With a great dock," Elisabeth interposed, leaning toward him and gesturing with her paintbrush. "And that's important, too—very important—almost more important than the house!"

"Well, now," he objected mildly, "now I wouldn't quite say . . . but it is important. No doubt about it."

"Maybe not quite that important," she conceded grudgingly, "but almost! But *this house,*" she declared fiercely, "if it comes to that, this is really some house, when you have a real look at it. Maybe it isn't any flaming beauty or anything, but . . ."

"I know," he admitted, trying to calm her. "I understand how you feel, but we've got to be realistic, now, don't we?"

"Right," I agreed, leaning forward, "realistic. What do you think?"

"Realistically . . . realistically . . . I think a fair price would be . . ." I hunched further toward him, as if he were about to make an offer himself . . . "would be about thirteen."

"Oh!" sighed Elisabeth, but she didn't fall. "Oh!" Long pause. "You do mean thousand, don't you? Or . . . or . . . or do you? Let's see. That would leave us with . . ." Her mathematics failed her as she hung from the post. "How much would that leave us with to pay, Bob?"

"Now, take *fifteen,*" he went on. "Fifteen would be a good price."

"A very good price," I agreed with warmth. "Very good!"

"Oh *no!* No, Bob. Not *fif*teen! We couldn't. We only paid twenty-two hundred."

"How long ago was that?" Ken asked.

"Thirteen years," I answered.

"Well, since then, land around here has skyrocketed. And besides, I bet you've put plenty into it, haven't you?"

"I'll say we have," she asserted. "We've put in plenty."

"If you had kept track," Ken said, "I bet . . ."

"But can we . . . do you think we can get thirteen?" I asked.

"If you wait long enough," he said placidly, "you certainly can."

"Long enough might be too long," I said gloomily. Then another thought struck me. "Suppose we sold it privately . . . without your help, I mean. Suppose some friend turns up and buys it? We've been advertising it, you see, and someone could turn up. . . ."

"Tell you what," he answered, "if you've got any prospects right now, give me the names. If any one of them buys it, then I don't get a commission. Okay? Is that fair enough?"

"Fair enough," I said hastily, and mentioned the names of four or five long shots, there being no short ones. He rose, and we followed him to the dock where his boat was tied.

"But," said Elisabeth nervously, "suppose we get only ten for it?"

"That's not enough," he replied.

"Maybe it isn't," she said, "but suppose . . . just suppose that's all we're offered."

"Well," he said, getting into his boat, "you'd either take it or leave it."

"That's all very well," she said to me, gripping my arm tightly, "that's all very well, but suppose that's all they offer? What would we do?"

"See you soon," Ken called, and cast off.

"But what would we actually do?" she insisted.

"See you," I called after Ken. Then, turning to her, "We'd leave it, that's what we'd do."

"But we couldn't, could we? Not if anyone actually offered it, we couldn't."

"We would," I answered firmly.

"But we couldn't!" Soon she had remounted the railing and was angrily swinging her brush.

"So what good was all that?" she wanted to know. "Paying him all that great *commission!* And what for? What for?"

"Give him a chance," I soothed. "He's not even back at his office yet."

"Well," she said, calming slightly, "you've got to admit it is nerve-racking."

It was.

For some people, selling is infinitely harder than buying. Why this is, quite escapes me. There is, of course, a very real difference between offering someone else your money and asking him for his. But why I should feel embarrassment to the point of shame when asking for his, when nobody seems to be the slightest bit concerned when asking for mine—it's most perplexing. Why don't other people suffer from my weakness? It had been painful to talk figures with Ken Keyes, and he wasn't even proposing to buy our cottage. How do those other people manage? That's what I want to know. Elisabeth is fond of explaining that it's all my fault because, she says, I never actually went to sell anything—anything at all, and in this, as in a few other things, she is at least half right. A long way back—I suppose a very long way—an industrious set of pack rats seem to have had a good deal to do with my origin.

Right now, for example, on the back porch stands more than one can of bent and rusty nails. It's not that I absolutely refuse to part with those nails, but the market just now isn't very brisk in bent nails. Even if it were, exactly how would I go about getting my specimens from so clear and definite a spot in space and time as my back porch out onto something so indistinct as "the market"? But even that achieved, what could I, what should I ask? A fair market price, obviously. But how is that determined? By what others are asking for bent nails? How do they know what to ask? From their point of view, I must be one of those "others" in regard to whose asking price they will arrive at theirs, but my trouble is that I haven't got the foggiest notion what to ask. To tell the truth, I suspect they're in no better shape than I am on this point. So where does all that leave us?

No. All things considered, the best, the simplest course seems to be to keep those bent nails, the pails of plumbing fixtures, ancient outboard motors, and so on. And it's not, I keep telling Elisabeth, because I am a miser. It's merely that I'm incompetent at grasping the subtleties of modern economics. I am fit only to muddle and bumble along on the island. But even that's no answer, because even into that innocent, pastoral life economics drives it pitiless and divisive wedge.

For example, the fact was that we had to sell the old place. But another fact, equally substantial, was that I didn't want to sell it any more than Miranda did. Though not precisely forming a contradiction, these contrasting facts created pressures, pressures complicated further by the belief of fear that no one would ever actually turn up to make an offer. After all, no one ever had before. Anyway, if they had, we would most certainly have said no and sent them packing. But on the other hand, if no one turned up now, we would most certainly be in a very unpleasant pickle. And if they did turn up, those strangers would tramp around most irreverently, perhaps even lounge—why, they might have the gall to sunbathe on our dock. They would probably go thumping around most disrespectfully on my planks without an ounce of feeling for those spikes I had driven. They might even cast disparaging glances at the spots where the hammer had missed the head of a spike and left its imprint in the soft cedar. They would be ignorant, very ignorant! I would have to explain very carefully that, though soft, two-inch cedar topping would long outlast the equivalent thickness of oak. To confess the truth, there was a time when I hadn't known that myself, a time when I wouldn't, in fact, have believed it if told by an eager seller. I would have been more than suspicious. These people, these prospective buyers would be equally suspicious. No. To hell with them! Let them go buy somebody else's cottage, but then . . .

In this happy and expansive frame of mind, I welcomed Ken Keyes with his first candidates, a middle-aged couple. The introductions completed, he took them on their tour, Elisabeth flittering along with them and dragging me.

"Nice dining room here," Ken remarked. "Overlooks the river. Beautiful view, eh?" No reply. "Could, all of you, sit right there and look out. . . . That table expands, doesn't it, Dr. Bob?" Being "doctored" always annoyed me, but Elisabeth sprang into the breach.

"Oh yes! Here, Bob, you take that end." She rushed to the other and began to pull strenuously.

"That's all right, Mrs. Russell," Ken said, "never mind. And now here's your kitchen. Nice gas stove there. Sink."

"What's that under the sink?" inquired the husband.

"That! under the sink!" I growled to myself, recalling the grim details of installations. It was a five-gallon hot-water heater from the Eaton catalogue. After putting in a set of mixing taps and adapting them for plastic pipe, I had crawled under the sink and had lain for days in pools of water continually replenished by leaks from above. In the end, though, it *did* work! It really did. There was hot water, very hot, for the dishes. Excited by success, I had even run it upstairs to the sink in the bathroom, where I exultantly shaved with hot water every morning.

"Small," observed the wife in a lackluster voice.

"It's more than enough for the dishes," Elisabeth defended with heat, "and for washing upstairs, too. Bob piped it up there as well."

"I meant the kitchen's small," explained the woman flatly.

"Oh, you should have seen it at the beginning," Elisabeth exclaimed. "It was much, much smaller. Originally there was a narrow pantry along that whole side, but Bob knocked the wall out, and now . . . why, now we can all eat here very comfortably." She had reminded me of the terror in my heart that Saturday night when I had sat in the smaller kitchen, ax in hand, praying for the courage to strike the first blow. Four times I had risen, strong with purpose. Three times I had brought the ax back over my shoulder. Twice I had touched the wall with the heavy blade. Then, the fifth time. The sickening crunch! The silence as I had waited for the timbers supporting the second story to creak, groan, and come crashing down. They hadn't. So, blow after blow. The ripping off of boards, smashing out the studs, and still no disaster. And next morning, that larger, lovelier, freer, and, Elisabeth said, lighter kitchen where we had all had room for breakfast! And that woman, that woman had said . . .

"Yes," the husband put in. "I can see where the studs came out right along there." What kind of a remark was that! You don't just knock out a wall and expect everything to look as if you had done nothing. You can't conceal . . . every sign, not every blasted one. The way he talked, it sounded like I had had nothing else to do besides shoving putty into nail holes in the ceiling. There had been docks to be built, hot-water heaters, porches. By God! If he bought this place, he'd find out soon enough!

"Good lot of shelves knocked up over over there in the corner," Ken tried to say with admiration. Knocked up! My thumb tingled with the memory of its suffering. Just let that guy try nailing angle supports into a brick chimney for shelves!

"Could do with a coat of paint," Ken conceded, perhaps watching their faces, "but that's a good thing, really. When you do the painting yourselves, you get the color you want, eh?"

"Mm," muttered the husband noncommittally.

"Shall we go upstairs?" Ken invited cheerfully. They shuffled off. I stayed in the kitchen patting my solid shelves. As they trooped and muttered around overhead, I could hear Elisabeth's bright explanations in the pauses. They thumped back down.

"We'll just have a little look round outside, Dr. Bob. We won't bother you again." Elisabeth went along with them.

"Have a good look at the dock," I threw after them with as much good humor as I could muster. Not many trees, I heard the wife saying.

"We planted all these!" Elisabeth expostulated. We had indeed. Dug holes for each one of those blasted things, and every one big enough and deep enough to bury a cow, and through the matted cables of turf, roots, and the stubborn clay beneath. My back ached just thinking about it. We had coddled each of the five trees through the arborial equivalent of chicken pox and measles. But no! they weren't enough to suit that woman! Not her!

When they had gone, Elisabeth tried simmering me down.

"But, darling, you've got to remember what Ken Keyes said. We've got to be realistic! And he's right. That hot-water heater under the sink sitting in the cut-down potty chair . . . it really does look a little odd."

"That heater's oval!" I snapped, "and the instructions said it had to stand upright! You try standing an egg on its end! Go ahead and try and see how far you get. The hole in that potty chair is perfect, and the arms help to hold the heater steady. As if it were made for the job. And besides, it's high enough off the floor to get a wrench under there to undo the plug for draining in the fall. It's perfect, just perfect! I don't care what he thinks!"

"But it does look odd," she declared emphatically, thumping my chest with a finger at each word, "it really does. Not to us. We're used to it! I quite agree the chair makes a fine stand for the heater, but people . . . strangers . . . they don't understand. To them it looks like we're . . . it just looks odd, that's all."

"I don't care if it does," I argued.

"You'd better care if you want to get any thirteen thousand for this place."

"Fifteen," I corrected.

"If anybody offers ten, we'd better grab it! That's all I can say . . . with a potty chair under the kitchen sink and everything." It wasn't all she could say, because she went on and on with the same theme for a long time. Ken's first couple, though, didn't give us a chance to grab anything. Maybe they did think the place looked odd. Maybe they thought *we* were odd, but I didn't care—them and their small kitchen, not enough trees, and so forth. Of course, when I thought it over, I began to care a little, but not much. Elisabeth seized her paintbrush, leaped back onto the railing, and returned to smartening up the place, but I refused to do a thing about the potty chair.

"We hear something about you selling," John commented casually as he dropped into a chair on the porch one afternoon. John was high up in a large chemical firm, and he and his wife Ruth had been spending their vacations on the island for the last ten years. Like so many others, after their first vacation they had been hooked, and came every year to fish and swim and generally laze through their two or three weeks. Our dogs had taken us beyond the level of mere acquaintance by the instant dislike they had taken to each other and the growling, howling battles that invariably began as soon as Pogo set foot on the island. After I had apologized two or three times to John and Ruth, and they had apologized two or three times to us, we knew one another well enough to discuss more interesting things over a cold ale and had become friends.

"Selling, borrowing, buying," I replied, "just generally doing my part to keep the cash in circulation in the good old American way." Then a thought struck. "Not interested, are you?"

"Oh," he threw off casually, "in a sort of halfhearted way. More curious than anything else." It developed that Ruth was curious too. She was so curious that she wouldn't even object to a tour.

"But they know the place already," exclaimed Elisabeth after they had left. "What do they want to look around for?"

"It's one thing to know a place as a visitor, but you look at it differently if you're thinking of buying," I explained patiently.

"We don't do that," she said. "How come we don't really go over things carefully before buying?"

"Next time we will," I said.

"Anyway, do you think they're really interested?"

"Hope so. They're islanders and won't ask stupid questions, anyway."

After their tour, they were interested enough to have a couple of carpenters look the place over and give their opinions. Then John paid us a business call. After much preamble and apology, he offered ninety-five hundred and withdrew.

"But," Elisabeth whispered angrily, thumping my arm, "you shouldn't have just said 'No' like that! It's an offer, a real offer, and we've never had an offer, not for anything, anything at all. Think! Think how that ninety-five hundred would help with the loan."

"You never take first offers," I explained.

"What do you mean you don't? Who says so? I jolly well would."

"That's because you don't understand business," I pointed out.

"I don't understand business . . . you think *I* don't understand business! What do you know about business, I'd like to know? What you should do is you should trot right down there and say you've thought it over, and we'll take it, and thank you very much. That's what *you* should do."

"Nope. I'm not going to." She went off muttering with her paint brush, to return at intervals with brief and spirited lectures about "being realistic, like Ken said."

Not long afterward, Ken appeared with more prospects in tow, and followed up their visit by telephoning their verbal offer of eleven.

"What did I tell you? What did I say?" I triumphed. "There's business for you!"

"Take it!" she gasped.

"I'll do no such thing," I declared firmly. "John asked me to keep him informed, and that's exactly what I'm going to do. This is the way business works, you see."

"The only business I know is that bank loan, and you'll have us mortgaged for the rest of our lives what with your turning down offers, that's what you'll do."

"Dad wheeling and dealing?" Richie inquired with amusement.

"Again!" she exploded. "Still! He's dotty, that's what he is, just plain dotty."

"Reply is beneath me," I replied, and I informed John and Ruth. In a day or so he came up with eleven-five, and Elisabeth wailed through my chuckles: "Now! Now will you, for God's sake, will you take it and forget all this . . . this *business*?"

"Now, now what we do . . . what *I* do," I announced carefully, "is I inform Ken's people about the exciting new developments!"

"But that," she objected, "that's . . . why that's like an auction, and you can't do that!"

"Business," I replied simply. "Besides, we still haven't hit that fair price of thirteen, so relax. Maybe, you know," I reflected, "maybe I *will* go into business after all. This is kind of fun."

"Atta boy!" cheered Richie and Mark while Elisabeth wrung her hands. Every time the phone rang, Elisabeth shouted to the children, "Never mind! That's probably for your father . . . your father's business," and they all huddled silently around me as I spoke cheerfully into the receiver, managing our affairs.

"Well? Come on! What did they say? What did they *say?* Come on! What's the market *today?"* Through the tense silence, I carefully explained the latest development, like Carnegie talking to his board of directors.

John and Ruth finally won the contest at twelve-nine, and Elisabeth put down her paintbrush with trembling fingers.

"Thank God . . . thank God that's over," she murmured. Then, with a trace of reviving annoyance, "but I bet you didn't think . . . you and your *business* . . . I bet you didn't give John and Ruth's name to Ken as one of those prospects, did you?" I admitted I hadn't. "So we've got to pay him his great fat commission, I suppose, and he didn't even really sell it."

"But he supplied the other bidders," I observed, "and that has to be worth something. Let's see . . . worth about thirty-four hundred," my calculations now having increased their speed enough so as reasonably to be called lightning.

The next day we gave our last loving touches to the old place. I was cleaning out the hearth after having burned the last stack of rubbish, and Elisabeth was clattering around in the already empty-sounding upstairs, recleaning, no doubt, what she had just finished scrubbing.

"Bob!" she called, "Bob! The darned toilet's stopped up."

"Plunger's there in the corner," I replied, picking up and starting for the door with my bucket of ashes.

"Done that, but it still won't work. You'll have to come." With a sigh I put down the bucket and went up and worked the plunger, but she was right. Thirteen years earlier that sullen bathroom fixture had presented me with one of my first island problems, and now it was threatening to bid me good-by with another.

"You don't suppose," she gloomed, "you don't suppose there's something wrong . . . wrong with the cesspool." She had uttered my worst thought.

"No, no," I reassured hollowly. "How could there be anything wrong with that?"

"There was once," she reminded me sternly, and we trudged silently downstairs and around to the back of the house. "That pipe . . . the pipe you and Norman put in . . . could be broken again. You said it wasn't exactly the right sort of pipe at the time. Remember?" I remembered, but that couldn't be the trouble. "How about if I call Norman? Oh, Norman!" she cried facetiously toward the cottage he used to rent, but he was off in Toronto professing the Middle Ages. Even had he been nearby, it would have taken more than a few of his cheerful "goddamns" to dispel the gloom that a little investigation revealed.

"The trouble is," I said, my voice trembling with deep emotion, "it's not the pipe this time. This time . . . this time it's the cesspool . . . the cesspool itself. It's just plain *full!*"

"But . . . but how *can* it be?" she wailed. "We've just sold it!"

"There wasn't even room," I soliloquized, "not even room for that last flush!"

"If it had . . . if it had only just waited!" she beat the shovel with her fist, "just that one last flush. If it had waited just that long, we could have gone, left, handed everything over and not known! Really not known. And then . . ."

"And then it would have been their problem," I finished.

"Exactly!"

"But . . . but . . ." I heard my better self stammering, "but now . . ."

"I know we couldn't," she agreed.

"Who," I said, angrily pounding her shoulder, "who was the one? Who took that last *flush?*"

She studiously ignored me with: "But . . . but who would have thought it would only take one! Only that one single . . ." and then she thumped me. "I suppose that's going to mean a monstrous great septic tank and a huge bill. The Public Health authorities won't let you fix up cesspools any more! When they go bad, they require you to install a septic tank! And, oh dear! there's Ken Keyes's fat commission, and now . . . and now this!"

"It's revenge," said Mark, "that's what it is. It's re*venge*. It didn't want to be sold."

"Told you," piped Miranda. "I told you we shouldn't have done it!"

"Man's life is brief and full of sorrow," I solemnized.

"But why are so many of them connected with plumbing?" Elisabeth wanted to know. "Richard! James! Why don't one of you take up plumbing? You'd be certain of a constant demand."

"And they get real money, plumbers do," put in Mark.

"Not me," said James. "I'm going to be a historian."

"But they're the biggest plumbers of them all," I pointed out. "All they deal with is the leavings of whole civilizations!"

"Is that the reason Norman is such a good plumber?" Richie laughed.

"And a good historian," I said. "Only difference is that historians are safe. The people they write about are dead and can't get revenge—not like toilets, anyway."

14

The Boathouse

"You can tell that Mr. Russell," came the message along to me from a regal island lady famous throughout the region for neither visiting nor being visited, "you can tell him he's going to have trouble with his boathouse." When it reached me, her remark seemed a little amusing. She owned a large island nearby, from which, as far as I had heard, she never stirred from mid-June, when she arrived, until her departure on Labor Day. How could she possibly know anything about the condition of the Mohawk boathouse? I knew the floor inside the door on the landward side was a bit bouncy, and I strongly suspected this was more than she knew. Whiz had admitted a little work there would be wise, but that only involved "shoving under a timber." Of course, this was easier to talk about than do. We would get around to it in good time. Nobody could do everything at once—not even Durrell. After the cabins had been cleared out and generally arranged, the first thing I meant to do was to pull off that damned sign on the seaward wall of the boathouse that declared us to be a girls' camp. The Russells owned Mohawk now.

While I hadn't the gall to advertise this fact, I certainly wasn't going to put up with being thought of by every passing boat as some holiday retreat. Pitiably innocent, then, of how one little thing leads inescapably to another, and yet another, I seized hammer and wrecking bar and started toward the boathouse to take care of this little job, this assertion of new ownership. I used to even think it was funny, that story about the woman who began by clearing off the top of the refrigerator and ended by having the whole house redecorated.

"Where you headed?" asked Mark. "What are you going to do?"

I explained.

"You ever been on that dock along the outside there?" I said I hadn't. He said he'd better come along because it wasn't in the best of shape.

As he stepped through the door ahead of me out onto the catwalk that ran along the building for forty feet, there was a brittle snap as of cracking sheet ice. He grabbed the door post and stepped back inside.

"A rotten board?" I asked.

"Whole thing's rotten," was his reply. "We're going to have to be awful careful." He tried again, and I followed. Together, walking on eggs, he guided me over and around the spots where boards had already actually disintegrated. Slowly we made our way along the wall of the building to where the six-foot sign was securely nailed. As I pried nervously under the heavy sheet of plywood with my bar, the topping beneath me creaked and whispered its dry warnings. It wasn't the fear of crashing through the dock that made us nervous. That would only mean a ducking in the deep water below us. It was the prospect of being raked on the way down by the rusty nails and spikes sticking out from the stringers and cribs beneath, or having, perhaps, only one leg go through and being pitched over, twisting that leg between the boards and along those waiting fangs.

"Look out, Dad!" Mark cried tensely, "it's going to come *down!*"

"That's what we want," I began almost angrily, but ending in genuine doubt. The sign was big; it was heavy. It, too, would be studded with nails; and it would be coming down on us, adding, if nothing else, its weight to ours on those crisply complaining boards on which we stood.

"Well . . ." he said doubtfully.

"We can't stop now. It's *coming!*"

"Look out!" we cried in unison, and it came, came with a crash, splintering the topping between us and the wall, but no nails.

"Success!" Mark laughed in disbelief, "and it didn't even kill us." Cautiously we picked our way back to the door, dragging the heavy sign behind us.

"How'd you get on?" Elisabeth inquired casually when I returned to the Main House.

"Sign's gone," I answered. "No more false advertising. Just a blank wall same as everybody else now, but the dock out there's rotten—all the topping's got to come off. Somebody's going to step out there, go through, and break a leg or something." The kettle was boiling, so I got a cup and sat down at the table.

"Mm," she replied, only half listening. "Too bad. Not hard, though, is it, just pulling up boards." She poured the boiling water into the teapot and got the milk.

"No," I conceded grudgingly, "not actually hard."

"Well then, what's the matter?" she asked, stopping in the act of pouring the tea. "If it isn't hard, what's the matter? If it's money, we don't have to put new topping on right away, do we?"

"No, we don't have to. It's just that . . . well, like Durrell says, *nothing's* easy, nothing."

"What's so hard about pulling up a few rotten boards?" She gave me my cup. I took a sip and let its genial warmth spread before beginning.

"All right. Listen. A rotten board three-and-a-half inches wide, four feet long, nailed with two nails in three different places—pull it right up with a wrecking bar and a hammer—no problem. So the board breaks, some nails pull through and stay sticking up out of the stringers, but that's no problem, only a nuisance. Still, there's forty feet of this stuff with three boards to the foot. About a hundred-twenty or -thirty boards. Well, pulling up that many is a little more than a nuisance, but not what you'd call actually hard. But," I emphasized, "when you get down to the business of beginning, there's the question of which one to start with. The next . . . really not the next, but almost the same question is: what are you going to do with the boards *after* you pull them up? Where are you going to put them? You're sitting there with a chunk of crummy wood in your hand, and you can't just sit there forever. And this is going to happen maybe a hundred and fifty times, what with the broken ones; so you decide to make a pile—that's not hard. But even making the pile's sort of dangerous, just handling that many boards with rusty nails sticking out of them, but you've got to make it. Making the pile means walking around with the damn things, and walking, what's more, around on that already dangerous topping; so you walk as litle as possible. That means you throw them, and that's not particularly healthy either. I know the kids can make the pile, and I'll insist they put on shoes when they do it. They'll actually put on sneakers, which is better than nothing, but not much.

"Where should the pile be? The back end of the boathouse rests on the built-up shoal, and if there's any sensible place to make such a messy pile of junk, that's probably it. So does this mean you should start pulling from the back end of the dock? If you do, you'll have to throw those last boards about forty feet to reach the shoal."

"Then obviously you should start at the front end of the boathouse," Elisabeth declared.

"Sure. All right, but that means that whoever's carting this junk to the back end to make the pile has to make two trips for each load across almost forty feet of treacherous topping—so rotten, in fact, it's too dangerous to walk on, and that's why we've got to pull it off. And he makes half of these trips carrying armsful of rotten, splintery boards, each one with five or six rusty nails sticking out. So, all in all—unless you prefer 'all things considered,' it doesn't seem to me a happy prospect! Still, we probably ought to start at the front end, since the kids have had their tetanus shots. So I pull the first board and chuck it behind me."

"But you shouldn't throw it," she objected. "Someone could get hurt."

"They sure could," I admitted, "but if I don't chuck it back a few feet, how am I going to move backwards to take out the next board? And the next? *And* the next? I'd be always backing up onto my own pile of junk lying there with its rusty nails sticking out just waiting for me to kneel down or sit down."

"Well, Bob," she said with decision, "it sounds too dangerous, that's all; just too dangerous!"

"So," I continued, "I've got to throw them back there far enough to be out of my own way for a little while, long enough so that the kids can pick them up and take them to the pile at the back. So what I do is I tell them to stay well back . . . which they do . . . so far back they'll forget the job, start horsing around, start shoving one another off into the river or maybe go through the topping. In fact," I concluded, "that dock's so dangerous we've got to do something about it, and that's what makes it dangerous to do, if you see what I mean."

"Good Lord," she said. "We'll have to hire someone; that's what we should do."

"Can't afford to," I said. "No. We'll do it, but all I've been trying to point out to you is that nothing's easy, that's all. You wanted to know why I was concerned about yanking up a few rotten boards, and I've been explaining." She took this in silence. "Then, of course," I went on, "there's the question of what's underneath."

"Underneath what? You mean the topping? I thought you said you called those long things stringers. Isn't that what you call them?"

"Right, but I'm wondering what kind of shape they're in."

"You think they could be rotten too?" she said with mounting anxiety.

"Why not? They've been sitting there for fifty years. They've got a right to be a bit punky."

"I suppose you could pull them off too, couldn't you?"

"That's right."

"But," she asked after a pause, "what's under the stringers? What holds *them* up?"

"Cribs. They're the foundations the whole boathouse rests on." (A crib is simply a huge box of heavy timbers filled with rocks. This is the standard foundation for docks or boathouses up there.) "And suppose . . ."

"You don't think," she said, catching on, "you don't really imagine that those cribs could be . . . be rotten! You don't really think that, do you? But if they were . . ."

"If they *are,*" I shifted her tense, "then we're really in the soup."

"Could we lose the boathouse . . . the whole boathouse?"

"We sure could."

"But that's a beautiful, a really beautiful building. We *couldn't.*"

"Relax," I laughed. "We haven't—not yet, anyway. We haven't even got the topping off yet, and that's the first step."

"But we've only just bought it! It's not even paid for yet and we're going to lose what . . . won't be paid for, not for years and years!"

"We haven't lost anything yet," I urged, "but we've got to get started."

"Oh, *don't!* Let's not! It's too dangerous!"

"Too dangerous if we don't," I countered. So next day we began.

The boards came up so easily, I had to concentrate on how nice the weather was so as not to think about the almost certainly rotten stringers.

"Hey, Dad," called James, who was helping to pile the junk, "here comes a tour boat. Let's listen to what they say about us now that the sign's down!"

As he drew level with us, the captain throttled back and the engines died from a groan to a soft rumble.

"We are now entering the Admiralty Group," began the loudspeaker. "On your right is Hay Island. At this end is a girls' camp run by the Kiwanis Club of Kingston, Ontario, and they take two hundred new girls every other week. On our left is Polaris. . . ."

"Dummy!" shouted James. Angrily, I rose, waving my hammer and wrecking bar, trying to look as unlike a girls' camp as possible, but it didn't help. After five years they haven't changed what must be their tape recording.

Anyway, what difference does it make what the passengers on a tour boat are told about who owns what? It doesn't matter to me any more, but at that time we had just bought it, just mortgaged our lives for decades, and it didn't seem fair for the tour boats not to take any notice of this fact. In what, I felt like asking, does ownership consist? Don't other people have to acknowledge it before it's true? Maybe I couldn't make the tour boats acknowledge it, but, by George, I could jolly well make the river folk—those idlers who wandered around in their outboards. I started worrying about them casually coming ashore. Not that they could actually do too much damage except with matches. Fire is a real and continuous danger on the islands, but even if they left all their matches at home, I didn't want them dragging their boats up my beach, disturbing my sand, maybe even rolling around some of my rocks, and generally carrying on in my underbrush. But how . . . exactly how was I to prevent it? We couldn't be everywhere at once—not always, anyway.

"Know what I'm going to do, Jack," I told Jack one afternoon when he had pulled up at the dock, opened the beers from his cooler and passed me one. "I'm going to get a whole pile of No Trespassing signs and put them up all over the show. Damned if I want everybody and his brother tramping around here. I come four hundred miles for privacy and quiet, and I mean to have it."

"And you've paid a pretty penny for it too, haven't you?" he added.

"Right," I said emphatically.

"You've got a right to do that then, I guess, haven't you," he said, thoughtfully lighting a cigarette. "Putting up those signs, I mean."

"I do and I will," I said.

"Do you ever listen to advice?" he inquired in a strangely tense voice.

"Sometimes," I admitted.

"Listen or not, I'm gonna give you some. *Don't.*"

"Don't what?"

"Don't put up no signs!" he said sharply. "Oh, you've got a right to, but don't."

"Why not?" I asked in surprise.

"Because it won't work, that's why not."

"Bet it will if I actually prosecute," I said.

"You do that and you're in real trouble . . . *real* trouble."

"I don't see why. Damn it! It's my . . ."

"It's yours, all right—that is, legally it's yours, and you can prosecute. Convict, fine . . . the whole business. A few other people have done that already."

"I bet it worked too," I said angrily.

"Yeah," he admitted, and after a pause added, "Then when they come back the next spring they found they'd had a fire over the winter." He stopped again to let that sink in. When it did, I got angrier, but it didn't bother him. *"Look!"* he went on emphatically and slowly, as if explaining to a child, "you're only a summer resident. That's all you would be if you'd paid a million for this. You come in the spring and go home in the fall. Right?"

"Okay, but . . ."

"But nothing! You're damn lucky I'm telling you, so listen! What I'm saying is that *we* don't. We live here . . . that means all year long . . . not just a few months. This is our home."

"All right," I argued, "but that doesn't mean you own this."

"What it means," he said carefully, "is that maybe I been shooting ducks out in your back bay here for twenty years, and I ain't bothering you when I do that because you're gone. And if I want to, I'll come ashore to drink a beer, and that ain't gonna hurt you either. Look! All the fellows around here . . . well, most of them . . . are pretty decent. You haven't had any break-ins or nothing in all the years you've been coming, have you?"

"No," I admitted. "Never had anything touched."

"And did you have No Trespassing signs up?"

"Nope."

"Then why start now? Why?" His voice was angry, urgent. "You start getting nasty about things like that, and so will somebody else. And when they start, mister," he warned, "you lose, and make no mistake about it. Unless you can figure out some way of taking this place back to Pennsylvania with you when you go, put it in a strongbox somewhere or like that, that's a game you can't win at. So just let things be. The fellows up here know you, and they sure as hell ain't gonna give you no trouble—that is, not unless you suddenly start interfering."

"Interfering!" I howled.

"Yes. Interfering. That's what I said. Maybe you did buy it, but you don't own it. Come to that, none of us do. Belongs more to the squirrels, the mice, rabbits, crows and so on—belongs more to them than it does to any of us. You want a squirrel to knock at your back door every time he wants to eat one of your acorns or something?" He began to laugh. "Jesus Christ! Whether you want 'em to or not, they won't, I can tell you that."

"Well . . ." I said sadly, "I don't know . . . maybe you're right, but . . ."

"No maybe's about it," he said. 'Don't none of us own much of anything, if it comes to that—nothing except maybe that six feet at the end somewhere, and then, sooner or later, somebody's going to come along and roust us out of that too." He laughed. " 'Course, maybe you're one of those guys that ain't going to die so as you won't never have to give this up, eh?"

"Maybe we're all only summer residents," I admitted grudgingly.

"Sounds like you might be smartening up," he answered. "But Jesus! You know, if I was worried like you about things I owned, I'd worry a little about that boathouse. She tips pretty bad toward the sea there. I noticed the other day when I was going by, you had taken off the topping. You had a look at them cribs yet?"

"I'm afraid to," I confessed.

"That's what I'd do if I was you."

"Damn it, Jack," I complained, "haven't you got any mercy? I can't think of anything I hate more than listening to a man who disagrees with me when he's right." Jack laughed and started his motor.

"I'd have a look at 'em anyways, and don't put up no signs. Get a mean dog if you want to, but no signs. Be seein' you," and he slipped away toward the channel, swung round, and zoomed off over his river. He left me, a summer resident, musing in the late afternoon sun about ownership. The years at the old place had taught me something about possession. What there was of it that was possible to achieve cost labor. I had better get on and have a look at those cribs.

They were rotten, all right. Some of the big timbers had come loose and floated off. Some of the rocks inside the cribs had already fallen out. More would. The foundations of the whole great structure were crumbling.

"Quite a place, isn't it? Yessir, quite a place," commented the fellow from Mitchell & Wilson whom they had sent to give me an estimate. "Be a shame to lose her."

"Yes. Yes it would," I agreed. "In fact, that's why I asked for an estimate . . . because I wouldn't want to lose her, that is."

"Yes. . . . I can see that. Will you look at all those saddle joints in the railing around the verandah up there! Lot o' work there. Every one of them done by hand and done well, too, from the looks of things. Really were craftsmen in those days."

"Sure were, but it's not the railing . . . it's the cribs—three of them—all along the outside I wanted you to look at." Unmoved, he stood on the dock and gazed.

"Cost you a pretty penny to put this up today it would, and that's a fact."

"Yes. They're around here, though," and I moved through the big door and started around the thirty-foot slipway toward the door at the back on the other side that opened onto what was now only a skeleton of spongy stringers. Reluctantly he followed.

"Don't you go out there," he cautioned, stepping past me and carefully out the door. He tightroped along the six-by-sixes, pausing to gaze down. Five minutes later he was back on the main dock admiring the saddle joints.

"What do you think?" I finally asked, breaking the silence.

"Beautiful place," he commented.

"Yeah," I agreed, "and it would be a shame to lose her, but that's why . . ."

"Tell you, Mr. Russell," he began, and I got ready for the shock, "tell you what it is. 'Tisn't easy to give an estimate. I can give you a figure, but wouldn't mean much. You see, there's no telling about places like this. You can't know what the job's going to amount to until you start tearing boards off, pulling things up. And you can't see what's underneath till you do that." I knew what he meant, but I wanted a figure to grieve over nonetheless.

"I understand that," I said, "and that's why I pulled off all those boards . . . just so you *could* see, and it's only those cribs . . . repair or maybe replace those cribs, that's all."

"Yessir," he assented. "The problem there, you see, is to lift this whole boathouse up, get the weight off those cribs, and you got to do that before you can either repair or replace them."

"Right. Yes. I understand that, and that's why I asked you to come and have a look . . . to give me an estimate."

"If you really want a figure," he patronized, "how about maybe five or ten thousand? But what does that mean? Easy to give figures." It *was* easy, as easy as saying thirty cents, but it wasn't actually that hard to know what it meant. I thought I knew that. So, year after year, she would just have to tip gently, farther and farther over, until, in the middle of some storm, she buckled and fell.

"But how, exactly," I asked, "how *would* you jack up something as big as this when it's sitting over nine or ten feet of water?"

"That *is* a problem, Mr. Russell, isn't it . . . a real one. Great place, though. Be a shame, an awful shame . . ."

"Thanks a lot," I said. "I know it isn't easy. Thanks a lot. You've given me something to think about anyway." And he had.

"Wouldn't be quite so bad," said Durrell, taking a breather on the front porch of the Caretaker's house that we had offered him the use of, "not quite so bad if that guy weren't so close to being right, and he is right."

"I know," I agreed, remembering Jack's lecture.

"We ought to give it some serious thought, though," Durrell said. "I'm kind of busy right now installing this dishwasher, shower, and whatnot; so that can be your job for a few days—thinking about it, I mean. Have you talked to Norman?" We had offered Norman and Betty the use of the Infirmary, and he was whaling away on it, cutting out walls, making doorways, and generally practicing on how to be a carpenter against the time when the bottom fell out of the market for medievalists.

"Yeah, but he's got his hands full swearing at the Infirmary. I'll think about it." I did, too—even came up with a few solutions, but every one of them disintegrated when laid out on the kitchen table and coldly scrutinized by the pitiless eye of practical reason.

"Come up with anything?" inquired Durrell one morning as he appeared at the Main House. Lots, I told him, and all bad. He admitted that, though his dishwasher, shower, and his whatnot were all working, he hadn't come up with anything practical on the boathouse himself.

"There's got to be, Robert Russell, there just has to be umpteen people around these islands who have faced this sort of problem, and some of them, by God, some of them must have figured it out. I don't know who they are, but I bet a couple of them drink beer at the International. Shall we try that?"

We did. We found one of them, too, and bought him a draught. He said it was simple. All you had to do was wait till there were twenty or thirty inches of ice. That much ice made a level, smooth, solid surface to jack from. In fact, that's when all the locals did this kind of work on their places. And besides, it gave people something to do in the winter. When I explained we weren't there in the winter, he admitted that that made things kind of difficult because it was harder to jack from water.

We had two problems: 1) finding a way to jack up the boathouse, and 2) figuring out how to repair or replace the cribs. They had to come in that order, too, because it would be no use figuring out how to fix the cribs if we couldn't get the weight off them. Despite that, we concentrated in our conferences on number two. At one of these, someone—maybe Elisabeth—said: "Why can't we sort of box in the old cribs . . . not take them out at all, but kind of build new ones around them?"

"That would work," said Durrell, "but the question is how to do it." As we were finishing breakfast next morning, he turned up with the answer to his own question and explained it there at the table, using silverware to demonstrate.

"Like Lincoln Logs," he began. "These cribs are ten-foot cubes. We take a twelve-foot timber and drill two holes in it eleven feet apart, and that leaves us six inches at either end. Then we do the same with another. Then we take an eleven-foot iron rod maybe a half-inch thick with a foot of thread for nuts at either end. We lay one drilled timber along the outside of the crib and the other on the inside . . . on the slipway side. We slide this rod through those holes we drilled, put on some big washers, then nuts. We do the same thing at the other end. Then we sink the timbers down as far as we want—and that's not going to be easy, because wood, that much wood, wants to float in the worst way—and then you bring in this friendly octopus whose hourly rates are low and get him to tighten those four nuts with a wrench. When they're really snugged up, what we'll have is a great big clamp holding things together down there with maybe a foot jutting out at either end. Well, then we sink an eleven-footer and position it so that it rests on these jutty-out ends, and spike that in place. We repeat the process till we build up to the surface, and we'll end up with, in effect, a whole new crib."

"And you mean . . . you mean," cried Elisabeth with excitement, "then we won't lose the boathouse after all!"

"Not for maybe another fifty years," he said.

"Oh, God, isn't that marvelous . . . just marvelous! But . . . but," with sudden sadness, "what was all the talk, then, about jacking things up?"

"We'll still have to do that," he said, "because you wouldn't want it sitting there caddi-wampus, tipping like it is now, toward the channel. But we'll get there, we'll get there because now we know what we're doing, and that helps. Oh, we have *details,*" he leaned forward with emphasis, "that have to be dealt with, but now we know what they are, and a little while ago we were in the dark."

"Details," she said gloomily. "Always those details!" He laughed.

"You're supposed to be cheerful now," I explained.

"Yes, but what are those details?" she mourned.

"Well, things like . . . you know . . . driving spikes under six or eight feet of water. It's hard to swing a hammer with authority under water," I said, "and these ought to be maybe ten-inch spikes. They'll take some driving."

"He's going to be a busy octopus," said Durrell, "but we'll get there. Somebody around here's driven spikes under water before. We've just got to find him and ask him how, that's all."

It wasn't at the International, either, but at a moderately fancy cocktail party that we met Ted Mabee.

"No problem," he said. "What you do is you take a length of pipe and a good solid rod that just barely fits inside the pipe. You put the end of this pipe down on the spot where you want to drive the spike. Doesn't matter how far under water, except for the length of the pipe and rod, but if the pipe's long enough, you position the end of it where you want the spike to go. Then you drop your spike down the pipe, then you drop in the rod, and whale the end of the rod with a sledge, and in she goes."

"By gosh!" I cried with admiration, "I bet that would work, too!"

"It does," Ted laughed. "I've got one of those rigs if you'd like to borrow it. Any other problems?" We explained about needing to jack up the whole structure. "Tell you what," he said. "I'll drop around tomorrow with the spiking outfit and have a look. There's got to be a way, you know. If this civilization can send people to the moon, jacking up a boathouse should be possible."

"Maybe," I admitted, "but if those fellows sat around waiting for me to figure out how to get them there, they'd have to wait a long time."

Ted was as good as his word, and his word was very good. He turned up the next day with his underwater spike-driver, took a few minutes to study the jacking problem, and then solved it.

"You make a T-beam out of two-by-tens," he explained. "Probably have to be sixteen feet long so as to go from the bottom all the way up to the top of that big doorway. But you nail a six- or seven-foot two-by-ten on the one end for a sort of plate to give the upright beam a solid base. Then just tack a little square across the other end of the T, just a little piece for a platform, and you put a hydraulic jack on the little platform. Position that right under the top of the door, and pump the jack. That's all you got to do."

"And you mean . . . you mean, then, this . . . this whole building will just go *up*?"

"Not the whole building," he laughed, "only this one end, but that'll be plenty good enough for you to fix up those cribs. Tell you what, though," he said sternly, tapping me on the shoulder, "Saturday is my inspection day, and I'm going to be around here a week from tomorrow. That means you fellows have only eight days, so get cracking, do you hear?"

We heard and cracked—in fact, we also bubbled, gasped, pounded, floundered, heaved, and some of us—mostly Norman—even swore for seven of those eight days. By the following Friday afternoon the boathouse sat on its new cribbing almost as squarely as it had fifty years earlier, and along that whole forty-foot outside wall stretched a lovely six-foot dock topped with two-inch lumber scrounged by dismantling tent platforms.

"The only thing I object to," said Durrell irritably, "is this pile of empty beer cases cluttering things up. Where'd they all come from, anyway? Makes you kind of wonder about the workmen. Anyway, I don't want evidence like that hanging around when the inspector turns up tomorrow. We might not even get any more jobs if he saw them, you know that? Still," he added thoughtfully, "a fellow almost deserves a quiet beer on the end of the dock, maybe even a cigarette, while he kind of looks things over." Norman opened, passed around cold ones from a fresh case, and took his out to the end of the dock where he turned to admire.

"You know," he said firmly, "goddamn it, Durrell, that is one beautiful job. It really is. Looks almost like a cathedral. You ever thought of building a medieval cathedral, Durrell?"

"If I ever decide to," Durrell answered, "I'm going to talk with Ted Mabee before I begin. Hope he gives us his seal of approval tomorrow."

"What I wish," I said, "is I wish he'd bring along that woman who said I was going to have trouble with this boathouse. I'd like her to see it now."

"She was a goddamn smart woman," Norman put in cheerfully. "She was right. You did have trouble. Not any more, though."

"What woman's that?" inquired Durrell. "I think I ought to know who she is if she's the one who started all this carry-on."

"As much as I'd like to," I said, "I can't really blame her."

"Then whom can we blame?" he pursued.

"I guess it was actually me," I admitted. "I just went down there with Mark one day. . . . We went down to take the sign off, and then. . . ."

15

On Ice

"No way," emphatically down the four hundred miles of telephone wire.

"But, damn it, Gordon, the geese have gone up! There must be open water or they'd have come back."

"You got to do everything a goose does? Maybe they've passed over down there, but they had enough sense not to come all the way to Gananoque. Stay where you are. It's only another few weeks before the season. Nobody! I'm telling you, *nobody* will go out on that river right now. Ice isn't thick enough to be safe, but too thick to get a boat through. Why, the Seaway's not even open yet. If they can't get ships up the river, how the hell are you going to manage?"

"Well . . . thanks anyway for the advice."

"Never mind the thanks. Take the advice," and he hung up.

"What did he say?" asked Elisabeth. I told her. "Mm," she commented.

At her insistence I had called Cuttle to find out about the ice because she kept interrupting our preparations for an Easter visit to declare firmly that she wasn't driving any four hundred miles to find out we couldn't get across to the island. "Because that would be plain silly," was the refrain, and then she would return to the packing.

Gordon Cuttle owned a marina, and his apartment was over

the showroom. He only had to look out the window to tell us what the river was like.

"You can never tell about the weather," I observed, stuffing my boots into a knapsack. "Take just a couple of nice days to break things up. It's only Monday, and we don't leave till Wednesday."

She hesitated, then went into her spiel, ending with "just plain silly." "James? James. Where's that woolly cap of yours?"

As fond as we had been of the old place, its call had never been strong enough to drag us from the comforts of central heating and the mild pleasures of watching the slow approach of a moderately southern spring. Mohawk was different. The voices of its trees, its rocks, spoke to the blood along the heart. They whispered under the clink of cocktail glasses, the general buzz of campus politics, the routine of homework, the spring-practice baseball games, birthdays, the buttering of breakfast toast. With a gradual crescendo, they rose as the days softened and the icicles dripped away from the eaves. Late one afternoon in early March, Richie burst into the house crying, "Hey! Geese! The geese!"

Chairs scraped back, dishes clattered into the sink, and everyone dashed into the back yard and strained toward the sky.

From behind and to the left came the sounds of distant argument—angry, insistent. Even from that distance, it was clear that each voice was determined to be heard. It had something to say, would say it, say it urgently, would be listened to, would be understood! As they came on, their argument rose till their shouts shattered the placid evening. Then they were over us—barking, yelping, a pack of hounds eager to frenzy, driving their quarry of winter before them. They gave tongue in that selfless energy of pursuit in which everything was expended—nothing held back—a dive more than a chase, a plunge at the throat of the pursued; no sense of danger for self, no fear, no suspicion of compromise, no hint of a dancing about out of range when the victim, in despair, turned to face the pack. Their mingled voices cried out the courage of life against death. Even the icy oblivion of winter itself cracked and gave way before the spear-thrust of geese, and they would not cease till they had driven him to earth above the Arctic Circle. There, on the borders of his last retreat, they would settle down, preen themselves, speak softly to one another, make love, and bring to birth incarnations of themselves that next spring would reconquer the bleakness of snow.

In the newly pregnant silence they left behind them, we stood trembling under the force of what they had said.

"Well," my voice rasped like the squeak of a chair in which a tense listener had just relaxed, "well," I squeaked again, "wasn't that something?"

"It's getting about that time," Elisabeth observed as we straggled back inside. We checked school vacations. There were five days early in April that we all had in common, so the matter was settled.

"But all the same," she declared sensibly, "we've got to find out about the ice."

I warmly agreed. Because it was sensible, and because I had agreed, but mostly because she kept repeating it, I actually did call Gordon, only to get the bad news that the ice was in that in-between stage—neither safe nor gone. We continued with our preparations nonetheless, and I promised to call Jack Stewart the next night to make more inquiries. I thought he was a good bet, since he didn't live right on the river and couldn't just look out his window to give us more bad news. Tuesday night I struggled through a kitchen cluttered with duffel bags and boots that wouldn't fit in duffel bags to make the call.

"Some channels are open," Jack said, "but no man in his right mind would go out there. Even the trappers have given up. Don't suppose it's any use telling you, though. I get out of work tomorrow at four. When'll I see you?"

"The International. Four-thirty," and I hung up.

Richie wanted to invite two of his friends, Mike and his brother Pat. We pointed out that it took a shoehorn to get the six of us into our VW bug plus luggage on top. He countered by reminding us that Mike had a microbus which, though it had a distinct taste for oil, did run, and eight plus luggage would be more comfortable in that than six in ours. We agreed, and, as if by magic, the two boys appeared with microbus, sleeping bags, and a few cans of oil. Early Wednesday morning we were lumbering along, battling a stiff head wind, with Mike at the wheel.

"But what, exactly, did Jack say?" Elisabeth repeated over the clatter of the engine.

"Just that it was chancy. Lot depends on the weather. I told you he might be able to rustle up an "in-and-outer" for us," and I re-explained about the little hulls with airplane propellers on the stern and sled runners underneath that could skim over ice and float across open water. She received my description of this strange craft as doubtfully as she had the night before. In fact, the more I described it, the less I believed it myself.

As we clattered along in the wake of the geese, her reports got darker and darker about the snow. North of Syracuse she gloomed, "Oneida Lake looks frozen . . . frozen solid . . . all the way. Not just in the bays and near the shore. Solid!"

Since I couldn't do much about Oneida Lake, I said nothing.

"Lots of ice down there," Mark observed, as we lumbered over the first span of the International Bridge toward Wellesley Island and Canadian Customs.

"Remember, everybody," Elisabeth instructed firmly, "remember that we're just going to visit friends for a few days."

"Aw, come on, Mom. Eight of us?" Richie scoffed. "Any customs man who'd believe that ought to be fired."

"Yes, but Richard, they're certainly not dumb enough to believe we're headed for a quiet weekend on an island in this weather. Use your head! We're suspicious enough as it is, what with all your long hair and with this . . . this . . . excuse me, Mike, old bus. We're really not all that respectable looking, are we?"

"What do you think they can do to us?" laughed Rich.

"Well," she said, "they can make things . . . uncomfortable. They could turn us out, search us, and I don't want to stand around in the cold while they go through all those duffel bags. And they don't repack them, you know."

"Oh, Mother!" Richie groaned in disgust.

"Well, they don't, Richard. In fact, we might seem more . . . well, more respectable if I were driving."

"You worry too much," growled Richie. "Anyway, here we are." The bus ground to a halt and Mike rolled down his window.

"And where are you folks headed?" inquired the official voice.

"Just visiting friends for a few days," Richie answered obediently.

A long pause. "Eight of you, eh?" Another pause. "Who are you visiting?" Embarrassed silence.

"Jack Stewart," I called. "Lives on Pine Street. Know him?"

"U.S. or Canadian citizens?" he said, ignoring my question.

"Neither," Elisabeth put in, leaning forward to show him her alien's registration card. He glanced at it and returned it.

"The rest of you?" he continued.

"U.S. citizens," I said in a loud American voice. He thought this over.

"You all one family?"

"Two families," I replied. "Well, one and part of another."

He considered this. "Step inside," he finally snapped.

"He means you, Bob," said Elisabeth, giving me a nervous jab in the ribs.

"Yeah. Come on, Dad," said Richie.

"I want the owner of the car too," declared the officer.

"That's me," said Mike. "Come on, Pat." The four of us climbed out and went into the office to lean against a long counter. Arming himself with official pencil and paper, the official officer began snapping official questions and writing official answers. "Name? Age? Date of birth? Wife's name? Age? Date of birth? Names of children? Ages? Dates of birth? Ever in Canada before?"

As the pencil hissed out my answers—or maybe official suspicions about them—I wondered why I was cooperating so meekly. In our fifteen years of border crossings, we had never had so chilly a reception. But was this a reception? Maybe it wasn't. Maybe in the end he was going to refuse us admission to Canada, and maybe his pencil was scratching out a record of those undesirables for whom his native land would have no room. As I wondered, the official turned to Mike and began the same treatment.

"For God's sake," I thought, "be polite—terribly polite." This fellow could, after all, refuse us. Four hundred miles, and then to be turned back! This was not a man asking these questions; this was an officer; he was power.

I searched myself for the courage to ask him just what the hell he thought he was doing, to tell him angrily that his government had calmly required and accepted taxes from me for more than a decade without even a word of thanks, that a friend of mine was a friend of Diefenbaker's, and that . . . that, goddamn it, what was this all about, anyway? I searched for this courage, and found it shivering behind the big, brassy reflection of practical wisdom that this officer was, in fact, *power,* and that, if I genuinely wanted to get to Gananoque, I had better keep my mouth firmly shut.

His interrogation over, he tore the official sheet from his official pad and said, "Would you come back here for a minute?" I wavered uncertainly, wondering how to get around the counter.

"Sure," was Richie's calm reply, and he disappeared.

"He wants Rich, Mr. Russell," said Mike.

"Oh," and I returned to make another stab at leaning casually. In our silence, I wondered, why Richie? Why not me? And that old answer I had heard in my mind over the years in similar situations came back. "You're blind, remember? Not a whole person. You're outside the pale, remember? Not a real person at all from his point of view—not one he can talk to in the ordinary way. He won't talk to you, not if he can talk to someone who can see." And the innumerable waitresses paraded through their years of diners and restaurants of memory with their question, "Does he want cream in his coffee?" Those waitresses with their surprise when I said, "He does." My twenty years of being a college professor would make no difference to the next waitress or the next customs official. Embarrassing, painful, but good to be reminded of a fact, when it was a fact. Trying to forget was no solution.

Richie returned, throwing a good-humored, "Yeah. Sure. Right," over his shoulder in answer to a mumble from somewhere far back in the big room. "Come on, Dad." We held our breath till we were outside.

"What was that all about?" asked Mike. Richie laughed . . . not loud, because half the laugh was working inside him with thoughts that didn't seem so funny.

"You won't believe this," he chuckled as we climbed back into the bus.

"Well?" was Elisabeth's tense query.

"You won't believe this," he continued, "but that guy wanted to know how we were going to support ourselves. He thought we were trying to sneak through to go on welfare."

Everyone laughed. "And when I told him Dad was a college professor, chairman, in fact, of an English Department, an author and all that, you should have seen his face! Come on. Let's get out of here before that guy thinks of something else." Stretched between fury and laughter, I held my tongue as the starter gave a weary groan, the engine spluttered, and we heaved ourselves back onto the road and across the magic line that defines the limit of responsibility beyond which the other government has to make the welfare payments.

"Now, children," Elisabeth explained as we drew into the parking lot of the International, "your father's going in there, and he's going to get us a boat . . . some sort of boat, and he's going to very, very quick, aren't you, because it's going to be dark soon. You are, aren't you?" she repeated with a solid jog in my ribs. "We'll be getting the groceries."

"Very," I said, slipping out the door.

"Five minutes late," cried Jack, clapping me on the shoulder as I stepped into the beery smoke and confusion of the men's bar. "Have a seat. Hey, Skippie! Two draughts to celebrate over here. Hey, fellas! Here's the first Yank, the first sucker of the season." There was a general laugh.

"Don't put calendars in Braille, eh?" Skip inquired after setting down the beers, shaking my hand, and making the change.

"They do," said Jack, "but he can't read that either! Hey, youse guys," he called to everyone, "any of youse got a boat in yet? Here's a crazy American wants to go out on the river." General laughter. "And with his wife and kids!" he added. A louder burst. "There's your answer," he went on. "Don't say I didn't warn you."

"Anybody got an in-and-outer?" I asked loudly. Soft buzz as people wondered.

"Walt Pullaw's got one," someone said, "but he's in Toronto this week, and ain't nobody 'd run that thing but him."

"Ain't nobody dumb enough," put in another. "Them things got no brakes!"

"I been trapping out on the islands for thirty-five years," said a river rat, "and I wouldn't go out on that ice there right now, not for nothin'." I got the message.

Though the International Hotel is an institution of respectable antiquity, I strongly suspect the Russells to be the only genuine family in captivity ever to spend a night there. Damn those geese, I thought as we climbed into bed.

Thursday morning, determined to take my medicine bravely, I walked into Cuttle's showroom. He greeted me with, "Damn it,

Russell, I told you not to come. Didn't I tell you not to come? Didn't I?" By this time he was thumping my chest with a big forefinger. "But you came anyway. You came anyway, and now you're here on my doorstep to rent a boat to get out there, aren't you?"

"Right, Gordon; absolutely right. Every word you say, sir, is true, sir."

After going through this routine four of five times, his self-righteousness was satisfied. Here again was power, and I cowered; but after each cower I added, "But what we need is a boat."

It was monotonous, but gradually he cooled down enough to look us over. Miranda, sniffling, her face flushed with fever, leaned against Elisabeth; while James, a cast on one leg, leaned on his crutch. Richie and Mark, Mike and Pat, all in sound health, wandered about examining the mixture of snowmobiles and outboards.

"And what happened to you?" Gordon asked James.

"Playing basketball in the attic in my socks. Slipped and broke my foot," was his laconic explanation. Gordon surveyed us in disbelief and dismay.

"Christ," he groaned to himself, "what a crew! You're a fine bunch to be heading out there onto that river—a fine bunch! Don't suppose you had the sense to get your phone connected, either." I admitted it sure would have been a good idea but . . .

"Figures," he sighed.

"But we'll take good care of the boat," I pressed, but submissively. You arrogant bastard, I thought. I didn't say it because that would only have made him mad, and besides, he was pretty close to being right on all counts.

"You're not going to believe this, but I'm actually more worried about *you,*" he growled, "well, not you, Russell, but your family here—I'm more worried about them than any boat." He sighed again. "Jeff!" he called. Jeff was sanding a hull leisurely and listening with amusement. "Let's have a look at these steel punts."

"They're frozen solid," Jeff observed, dropping his sandpaper and strolling toward the door. Gordon turned on us with his captain's air.

"Tell you what," he almost barked, "I'm in charge, see? I'm in charge."

"Yessir!" I returned with military enthusiasm.

"And don't get snotty, either, or you'll sit here till June. What I'm saying is this: I won't rent you a boat, not one boat; but, if we can break these punts loose, I'll rent you two." Businessman to the end, I thought. He's got me. He knows it, and he's going to put it to me till it hurts. "Now, you listen to me," he roared, as if reading my mind, "the point is that one boat—I don't care what boat it is—no single boat is safe out there right now. Motor conks out—anything like that happens—what the hell are you going to do? Get out and walk? There's just no help to be had out there. There's got to be two boats." Of course, he was right again. "So, if we can break them loose, you take two or none."

"I'll take both, Gordon," I agreed.

"But I go first," he continued emphatically. "Jeff and me, we'll go out and look things over . . . see if it's possible to get as far as Hay. Remember that it's me that makes the decision. Understand?" Whatever else he was, he wasn't misunderstandable.

The boys dove through the door to where the two punts lay upturned outside a storage shed and began chipping the ice away. Soon they had them free, turned, and sliding over the heavier ice by the shore. Then the double crash as they went through. We cheered. Not to be done out of any sport, Richie and Mark took their places in the bows and began breaking ice to make lanes to the open water. When they had made enough room, Gordon and Jeff put a ten-horse outboard on the stern of each punt, and when the boys had got through the heavier stuff, they started the motors and smashed the rest of the way out under power. We cheered again as the two droned off up the channel. Not till then did I relax enough to notice how the knife-edged wind slashed at my face and cut through my heavy coat. I ordered Miranda back inside before she caught pneumonia, if she hadn't already. She argued that from inside she wouldn't be able to see the boats and whether they could get through. I didn't answer because my attention was gradually absorbed by what must surely be one of the strangest and most beautiful sounds in the world.

As if breathed from flutes of crystal, eerie notes rose from the river into the frosty air—each different in pitch, each clear, cold, vibrant; they combined in discord to make the perfect chord, which resonated back and forth from far right to far left, and yet always in front. But then not a chord, really, so much as a ringing in the ears, as after a heavy blow. The ringing, though, was not inside, but outside, so perfect a sound that it seemed both outside and inside, a miraculous music that, by its perfection, united the self with that which is other than self. Reverently, we stood as the mystery played upon and through us, ourselves becoming part of its instrumentation.

"Ice!" cried Elisabeth. "It's the ice! The wind blowing over holes in the ice!" Though this must have been familiar music to the ears of countless generations in the north, we heard it then for the first time and our hearts had stopped. And never again would this music take us out of ourselves in that same way because it would be familiar—the word we apply to things from which experience has shorn the glory. Who has ever recaptured the ecstasy and pain of his first love? The wish to do so is strong enough in us all, strong enough to have built mansions for many a pornographer who lures us along with the promise of reliving that great discovery through vicarious sexual extravagance. Though success is never possible, neither can we stop hoping. But perhaps it's just as well. Maybe that dulling, that padding, that softening of experience, that becoming "familiar" is what makes a reasonably ordered life possible. If waves, wind, sun, smile, skin, voice, movement, touch,—if the spears of all or any of these were not blunted, which of us could survive a week in July without being blown apart by anguish, passion, and joy?

A middle-aged thought? But maybe God, when he got to be middle-aged, realized it would be a downright blessing to pour over and through these human creatures a large dose of the capacity of becoming used to things. After all, one can wriggle on the pinpoint of experience only so long. It would be diabolical to keep him forever squirming. No. He'd better be able to get used to things, and pretty quickly, too. So, with kind caring, God poured a liberal dose of thoughtless adaptability into our quivering clay, even into the Russell bit. As the seconds passed, I gradually became aware again of my surroundings as the power of that music faded. When we went naïvely back to urge Mrs. Cuttle to come out and listen, she said, "Oh, the ice! Yes. Does make funny noises, doesn't it? You ought to hear it *sometimes,*" and she went on working over her account books.

"Well," groaned Gordon on his return, "I suppose some lunatics might be able to make it over there. Why anybody 'd want to, though . . . that's another question, but I imagine I'm looking at just such a bunch." He stopped the sudden scramble and rush with, "Hey! One more thing, and I want everybody to listen. Do you all hear? *Listen.*" Dramatic pause. "When you get out onto that ice, any of you—and there's no way you won't, all of you, even you, James, with your crutch there—when you do, remember one thing. If you go through," he paused, "if you go through, you've got just thirty seconds . . . repeat, thirty seconds—of useful consciousness in front of you. Thirty seconds. No more. That's it. That's all she's going to write. So, if that happens, *move.* Everybody move! And that means nobody goes out on it alone. Stay close together . . . roped, if possible . . . but I suppose it's no use telling you to do that, but stay close and be careful. Just keep in mind that you're customers of Cuttle Marina, and therefore valuable; so, goddamn it, be careful!"

"Finished?" I asked.

"Finished," he said.

"Okay, boys!" I called. "Load up!"

As he strode through the bustle of loading, Gordon was muttering about the lame, the halt, and the blind and about his not being responsible for crazy people.

In a few minutes Mark yanked the starter cord and steered slowly out down the freshly cracked lane with me, James, Pat, and luggage aboard. I asked him what it had been like out there over toward Hay.

"No sweat," he said, twisting the throttle, bringing the wind across my face like an ax. After ten minutes of clenching against the cold, Mark cut the motor back to trolling speed, and the wind lost part of its edge and force.

"Can't get anywhere near Mohawk, though, Dad," Mark said. "Going to have to go in at the old place. There's ice there too, but I think we can smash our way in pretty close. Keep down and hang on. I'm going to ram her in," and he spun the throttle to full. We hit with a splintering crash. Recovering, I checked to see if the bow was still intact. It was. Thank God for steel, I whispered shakily.

"Dad?" asked Mark coolly, "can you shove us off with the oar? Got to make another run." I pushed us off. We backed out. He gave it the gun and we crashed in again. Then a third time. And still the steel held.

"Okay," said Pat. "It's thick enough here to get out," and before I could stop him, he did. He had dragged out the first duffel bag before Mark cut the motor. Then Richie's boat crashed into the edge of the ice behind and to our left. Gingerly, I crept over the bow and stepped onto that "thirty seconds" ice, trying to figure out whether one actually could hoist oneself up by one's own bootstraps. The ice held. I helped James out. He wasted no time trying to walk, and crawled swiftly away, dragging his crutch.

"Dock's over here, Dad," his voice called from twenty-five feet behind me. Crossing to where he crouched, I found the friendly old planking, but, the water having dropped before freezing, the dock was shoulder high. Hampered by my heavy clothes, I clambered up awkwardly, lay down, and hoisted James up. Mark and Pat unloaded and passed up the baggage from both boats. Miranda, who had come with Rich, struggled up beside me with a sniffle.

"Mike and I will go back for Mom and the rest of the stuff," called Richie.

"No," I shouted. "Both boats go."

"One's plenty," he called back. "There's not much more."

"Both boats!" I shouted. "Cuttle was right. No boat, whoever's in it, goes alone. Out here we go in tandem wherever we go."

"But it's cold," Mark objected. He was right, but that was why he had to go.

"Sorry," I said, "but both go." Reluctantly, both pushed off and started back over the mile and a half of wind-whipped liquid ice. "Think you can walk, James?" I inquired.

"Nope," was his sad reply. "Wish I could. This leg is so cold!" With a sudden shiver I realized I hadn't thought of that leg immobilized in a frigid cast. Bending down, I picked him up and set him on my shoulders.

"Now, your job, James, is to hang on," and I started down the slippery catwalk toward the shore with a duffel bag under each arm, Miranda and Pat following with their share. The wind had swept the dock almost clear, but there was snow on the land, knee deep and, in places, hip deep. Ten yards and I was bathed in sweat. That half mile up over the cliffs to Mohawk and the Main House—burdened this way, it was impossible. Dropping the bags, I headed for the front steps of the old place. Even to reach them would seem a triumph. Mohawk seemed farther away now than it had yesterday morning in Lancaster. When I finally made the steps, I put James down and sank gasping beside him. Though Pat was a gymnast on his high school team and in great physical shape, I could hear him puffing as he and Miranda struggled up. No doubt about it—we had a problem, and not one that could be mulled over carefully—not with James's leg; for that matter, Miranda was in no condition for a leisurely wander through the snow. We had to move, and not in easy stages. The question was whether I was able.

"They don't do much about sweeping the streets up here," Pat commented dryly.

"We've got to mush on all the same," I said, rising. "Don't worry about the baggage. The chief thing is to get there right now. We can come back for this stuff any time." I hoisted James to his perch, Miranda took my hand, and we plunged back into the snow. It would have been hard enough going even without a passenger. With one, as I floundered along, stumbling and slipping on the rocks and logs beneath the snow, my heart began to pound, then to hammer. My blood beating against the skin felt like some caged bird throwing himself against the bars in desperate efforts to escape. Almost awash in perspiration, I wanted to tear off the heavy coat that muffled and hampered my every movement. I needed the wind to cool down the boiling, dripping body inside. "And this," I thought foggily, "this is not the place, not the time for my heart attack."

Maybe Pat was reading my face because he said, "Okay, Mr. Russell. It's my turn. Come on, James. Climb aboard," and he whined and yipped like a husky anxious to take the trail. James gone, I wrestled my way out of the heavy, dead weight of coat and stood for a moment trembling with relief in the icy air. Feeling easier, we pressed on.

"Hey, Dad!" cried Miranda excitedly, having been struck still by a thought, "what about the sled? Didn't you say you found a sled up in the attic last summer? Didn't you?" Whether it is given to all of us to have moments of genius I'm not sure, but this was certainly one for her. But where had I left that sled? Had I taken it down? Surely not. What use could I have had for a sled in mid-July? No. It must still be in the attic. Attic! This was Pat's first time on the island, so he didn't even know where the house was, let alone the attic. Though Miranda knew, she wasn't tall enough to reach the trapdoor that led to it even from the top of the stepladder—assuming she could have found the stepladder. I groaned. Her moment of genius was offset by my inability to see how to put it to use.

"You're right. Good thing somebody around here is thinking. Maybe we can use it later, but I don't see how it can do us any good at the moment." So, doggedly, sullenly, we sweated, slithered, and floundered on, Pat and I taking turns carrying James.

"Hey," Miranda broke the silence again, still thinking, "how about going across the bay? Be lots quicker. Lots easier. The ice is all clear and we could go in a straight line."

"Only trouble is," I gasped, "only trouble with that is how thick . . . how thick is that ice?"

She could see my point, but added, "It would be quicker, though."

I agreed . . . might take only thirty seconds that way. We slogged on, resting only when another step seemed impossible. At last I sagged and leaned trembling against the back door. I took my time about reaching into my pocket and pulling out the keys because, for a while, I was shaking too hard to fit the key into the lock. Kneeling, I let James slide off, then opened the door. Making our way inside, we collapsed into the nearest chairs. After a long, exhausted silence, I asked, "How's the leg?" A silly question, but I asked it.

"Can't feel anything," he returned stoically. On trembling legs I rose, went to the little stove in the corner of the dining room where we sat, and began working on a fire. As the kindling began to crackle, I knelt and removed the heavy sock on his bad foot and started massaging it.

"It's the leg too," he admitted. "The leg is cold." But I couldn't get under the cast to rub. To save heat, Miranda pulled the heavy curtain across the doorway of the little room sealing us off from the huge kitchen. Slowly, degree by degree, it came. Leaving James with his leg propped on a chair, I returned to the back porch, found the stepladder, and hoisted myself into the attic. It was actually there—a real sled. Jubilant, I passed it down to the waiting Pat who took off with it, as I later learned, toward the bay. He had had too much sense to get my approval for what he intended. I would have said no if he had asked me. He knew that, so he didn't ask, and he proved to be right. The ice over the bay was plenty thick, and he fairly whizzed over and back with a load of luggage and groceries. When Elisabeth arrived breathless via the overland route, she found one warm room, two silent children, and one almost recovered husband.

"Cuttle was right," she panted, collapsing into a chair. "We *are* crazy—the lot of us." And, after a pause, "But it is marvelous to be here, all the same."

"Even though Cuttle says we're crazy," she panted, sinking into a chair, "we're not, you know. We're not. Coming up over those cliffs, it was so beautiful! It was all like that story, remember? about the princess who lived on a mountain of ice—everything glittering, flashing, sparkling . . . everywhere clean, dazzlingly white . . . and down there on the bay, there were Mark and Pat scooting along with the sled and the sun reflected from the ice . . . like they were streaking over a field of diamonds. If I could only explain! But the trees . . . all glistening and still, and in the snow the tracks of all sorts of little creatures! Wish I knew what they were . . . it's like a picture from fairyland, you know, except it's *real* . . . and those tracks! Things are alive here, sniffing around and nibbling whatever they nibble. . . . And it's *so* different from summer . . . so different, but just as beautiful . . . in a way, even more!"

She spoke of what her eyes had seen, and for her, that had been enough to refute Gordon, to refute any snug, secure mainlander. But for me, what she had described lay on the other side of the curtain. Why had I come? Why was it enough for me, though still shaking from exhaustion, simply to be there?

The geese, partly—what they had said, what I had heard through their speech about the fresh business afoot in the world and my desire to be there while it was happening. There? But where was there?

Wherever it was, it was surely not in Lancaster, not in any city. The city was its own world. It consisted of permanent, but empty abstractions: parties, programs, policies, corporations, reforms, controls, inflation, government. And it was no better on the campus, where departmental and collegiate politics flared over this issue, then that, and everyone was forever and always writing reports, long ones,—and stuffing them into my pigeonhole, demanding that I read them, that I make up my mind, and that I urge, argue, and cajole others into voting this way or that about black studies, yellow studies, urban studies, English composition, the need for a foreign language requirement, the absurdity of a foreign language requirement, my own absurdity, everybody's absurdity. And in the whole bloody, screaming, turbulent pack of them, not one single nit or grit of substance! *Abstractions,* the whole pile of them. And, make no mistake about it, they do have to be dealt with, those abstractions, but, after a six- or eight-month wrestle, it all begins to seem like a boiling madhouse of gibbering, struggling ghosts flailing empty arms through one another's vacuities.

But here on the island, life, or maybe the contest, seems real, the forces clear. Anyway, it is believable. When the sun shines and the breeze is soft, it is loveliness past speaking. The very pores of the skin become ears to drink in the sound of slapping waters. When the wind blows—and it can blow, blow strong enough to blaze the whitecaps right off the waves it has raised—then Danger walks the earth and with heavy tread. No doubt about it! It can be mortal, but life is mortal, and not just for some, either. Housed in foam-rubber, plush-lined steel cocoons and hurtled along smooth highways at seventy miles an hour, the only sound the subdued rushing hiss of air, thus sealed off, sealed up, hypnotized by that whole complex lie—who can believe in the crash, in the splintered bone, in the tattered flesh, the scattered teeth? Wind and water do not lie. Danger is real, as real as the sweetness of summerings. It is this truth that the wind speaks, that the waves thunder, that the rocks stubbornly affirm—this is the truth that first drew me to those islands, and it is this which will keep me there. Right or wrong, wise or foolish, it seems like truth to me, and in that "seems" lies my necessity.

Whether this or anything like this lies beneath the children's answers when we ask if they would like to go to the island, I do not pretend to know. They simply answer, "Canada? I'll say!" That's why we were all there.

The boys whisked the supplies across the bay on the sled and lugged them up to the kitchen where Elisabeth received and organized them simultaneously. At the same time, under her efficient hands the big, bottled-gas range in the kitchen began magically to transform the dead, barnlike atmosphere into the living warmth of a kitchen by sending forth the sounds and smells of frying meat, potatoes, and onions. Before long we were all wedged about the table in the small, warm dining room, the wood stove crackling and humming in the corner as we scraped our paper plates for the last shred. Then more hot tea to go with the "olde English fruitcake." Between the intermittent sighs of satisfaction that rose as the scraping subsided, I heard the deep, steady rushing of the wind outside rising slowly in pitch. Reaching down through the cloud of heat by the stove, I slid the draft bar shut.

"Wind's coming up," I observed. "Sure those boats are all right?"

"Yeah," said Rich, yawning. "We couldn't get them right up to the old dock . . . well, not all the way, but they're tied. We pulled them out onto the ice . . . well, not all the way out, but they're okay."

"Long as you're sure," I said, "because this isn't exactly the time to be here without transportation." He said he was sure and yawned again. We were all tired; so, after cleaning up, we got ready for bed. Shouldering luggage, Mark and Pat set off through the snow for Mark's cabin, and Rich and Mike headed for Richie's. I carried James over to the Master Cabin where he and Miranda shared a bedroom next to ours. Even after a good meal, and over a familiar path, it was heavy going through a foot of snow; the wind, though broken by a screen of trees, drove and tore. The bed, when we finally crawled in, was so cold that it was downright funny. Till then, I hadn't realized one can't laugh and shiver at the same time—not really shiver—and it was too cold to stop shivering so that one could laugh. The more one couldn't stop, the more ridiculous became the situation, so the need to laugh increased, but laughing wasn't possible because of the need to shiver. Terrible predicament, really. After what seemed a half hour of shivering so violent as to approach thrashing, we subsided into quiet as much from exhaustion as from satisfaction with the temperature of the bed clothes. Suddenly the door was flung open and Mark stomped in with the announcement that he and Pat were going to check the boats. I listened to the wind, whose voice had risen to the shriek of full gale, and thought this over. Both were strong, athletic boys. On the security of those boats depended our security.

"Got flashlights?"

"Yep."

"Then take along the heavy nylon line on the back porch to tie them with. Don't go over the bay. I'm still nervous about that ice. Stay on shore and be sure to check in here as soon as you get back. Okay?" He said okay and thumped out, shutting the door.

"You think they're going to be all right?" Elisabeth asked uncertainly. "That wind sounds really up!"

"Sure. They'll be all right." But I wasn't sure. Still, the boats really ought to be checked.

My watch said eight-thirty. Maybe, at the outside, an hour to go back to the old place, make the boats fast, and return. Listening to the voice of that big wind, I waited.

Nine . . . 9:15. . . . Any time now. . . . 9:20. I couldn't expect to hear them coming back over the noise of the gale, so I focused on the door. . . . 9:25. The only sound inside was Elisabeth's even breathing beside me. My ears fought to reach through the howling outside for the crunch of a boot on snow. Only the hissing roar. But it might . . . might take more than an hour. Probably it would, I lied. The snow was deep, though; very deep in places, and in the night . . . plus the wind. . . . 9:30. It couldn't possibly take an hour! What kind of father was I to let him, to let them, go? Why hadn't I gone with them? Boats! Boats be damned! We had our own in the boathouse with motors we could have unlimbered if necessary. As gently as I could, I slipped out of bed, into my clothes, and then out to meet the ax blow of the wind. I would try the Main House first. They might be there, huddled over the glowing stove trying to bring the blood back into what must be numb fingers. My feet could not find the summer-worn path beneath the snow, so I set my course by dead reckoning, checking my position by familiar trees. At last I reached the step outside the dining room, found the knob, and opened the door.

"Mark? Mark?" Only the crackling of the stove and the impersonal, resolute ticking of the clock in the room. "Just remember. Thirty seconds." I was remembering. I could make the boathouse . . . maybe I would meet them coming back—that is, if they disobeyed instructions and had tried to come back across the bay. Through the empty dining room, cold kitchen, the enclosed back porch, and out again into the stinging night.

Cautiously I moved in the direction the boathouse must be and the steps that brought one down the fifteen feet to the dock. Very cautiously, because if I missed the first steps, I might go over the edge of a twenty-foot drop to the rocks. A birch rail had been nailed up across the end of the path leading from the Main House to prevent precisely such a mishap, but if I strayed too far to the left, I could pass around the end of the rail—that is, if the rail was still in place. Bent forward against the heaving of the gale, each foot feeling carefully through the snow for solid ground, I crept along. Bump! It was the rail across my thighs. Luck was with me this far, then. Turning to the right, I found the three cement steps leading to the little bridge connecting the second story of the boathouse to the shore. Down those steps, across the bridge, then down the fourteen steps to the dock. Out to the end and then over the end onto the ice.

Now what? What, actually, had I come for? What had I really expected? Had I really thought I was going to meet them? No. Not really, but I had had to go as far as I could in the hope . . . just on the outside chance. This, the end of the dock—it was as far as I could go. Ten steps farther out onto the ice beyond the end of the dock into that echo-destroying blaze of wind (the blind navigate by echo) and I, too, would be lost. Calling for help would be as useful as telegraphing the dead-letter office. So, one hand clamped foolishly on the last plank, I shouted for my son. Again and again—louder and still louder, but it was only to shout into the thunder of Niagara. Had I really expected anyone ten yards away to hear, let alone a half mile? It was madness.

I swung back onto the dock, up the steps, across the bridge, then up again to the path for the Main House. They could have obeyed instructions, stayed off the ice, come back overland; maybe now, maybe now they were in there by the stove. They weren't. Then back to the Master Cabin and our bedroom.

Elisabeth was still asleep. Ten minutes to ten. Stepping outside was instantly to raise the question of one's survival. I had known that. And just for a couple of boats, I had said, "Okay. Go ahead," calming my conscience with a "Be careful." Not going myself, but saying to my son, "Go ahead." At fifteen and on familiar ground, how could two boys be expected to believe in death? I believed in it, but then, I was over forty.

Reaching out, I shook Elisabeth gently.

"I think we'd better go check," I said softly. "It's almost ten. They aren't back."

She moved slowly at first; then understood. Then up and beginning to dress quickly, not mentioning the cold. What, I wondered, what exactly would we, could we do if they weren't at the old place? Coast Guard? Call the Coast Guard? When in trouble, call the Coast Guard. No. In Canada it was the R.C.M.P., but how call without a phone? And even with a phone, what could the voice at the other end say on such a night except: "Thanks . . . thanks for the . . . report. We'll see you . . . we'll see you when . . . when the ice goes out." What else could anyone, Coast Guard on R.C.M.P., say?

"We'll need a flashlight," I said, because I had to say something.

"Okay. Ready," and started for the door. Before I reached the knob, the door was flung open, boots tramped in, and the door slammed.

"Wow! Wow! Some night!" said Mark. Churning, I retreated to the bed and sat down.

"Wind was so strong out there," Pat said, "I had to crawl, really crawl out onto the dock to the boats. Couldn't even stand up!"

"Lucky we went," continued Mark. "Both were taking lots of water over the stern. We dragged them up as far as we could and put two more lines on each. But wow! What a night!"

"I went down to our dock and called for you," I said.

He laughed good-humoredly.

"But you're all right? Both of you? You're really all right?" Elisabeth questioned earnestly.

"Oh yes, Mother dear! Yes yes, your son is all right," Mark crowed in falsetto with exaggerated emotion. "Your tiny boy is just fine!"

"All right, damn you," I half laughed, half cried, "go on over and warm up by the stove; then, for God's sake, get to bed!"

"It's too cold to go to bed," Mark objected through chattering teeth as he and Pat stamped out again. And moment went on following moment, just as they always had and as they would have even if I had not been sitting there, winding down, listening to the thud of their boots on the porch.

By morning the wind had dropped and, leaving James and Miranda to tend the stove, we made an expedition to see how the boats had fared during the night. Like two huge ice-cube trays without racks, they lay as if cemented in place. The gale had driven them askew, sliding them till every mooring line must have sung. Now they were iron rods. On each stern sparkled a glassy imitation of a ten-horse motor exquisitely modeled overnight by spray and wind and cold. For two hours we chipped away at the half-inch glaze on the motors, chopped great slabs of ice out of the boats themselves, and worked with numb fingers at the metal balls that had been knots in the lines. Miserable work, but if it hadn't been for Mark and Pat, we might have been looking for the boats instead of restoring them. Usable? Would those, could those motors possibly work when we eventually did unlimber them? Though my ears told me so, and though I was accustomed automatically to believe them, I felt sure they must be lying to me when, after a few yanks of the starting cords, the cylinders of both motors cracked; then, at the next pull, roared.

"Okay!" Richie cheered triumphantly, "we got transportation. Where shall we go?"

"Back for a hot cup of tea," I answered. He blew up.

"You mean to tell me we did this . . . all this, and now we're just going to tie them up again?" I explained that we had needed to know whether they would work—that it was even kind of important to know—but he wasn't impressed.

Confident now about the bay ice, we slid quickly home. James had been doing his job feeding the stove, and the dining room seemed actually hot. He sat in comparative luxury, his leg propped on another chair, Miranda sniffling quietly beside him. Sniffling herself as she poured the tea, Elisabeth observed that she was coming down with it too. To suggest they both take to their beds would have been like advising two pneumonia suspects to have a good rest in some nearby refrigerator; so for the next two days they held friendly communion with the stove, Miranda rereading for the third time Francis Hodgson Burnett's *The Secret Garden,* and Elisabeth correcting a large batch of essays from one of her classes in English literature—she was then teaching in the English Department of a college near Lancaster.

During those days the boys were desperate to get out onto the river—or, rather, the open water. I banned any cruising on the grounds that, when it came time to go back, we would be very grateful for the remaining gasoline in the tanks of the punts. I added that, if they could find an open gas station, they could go anywhere they liked. They couldn't manage that, but did the next best thing. Rummaging around in the boathouse, they came up with a tankful left over from last summer, so I gave permission to launch our big fiberglass. Easing it down over the end of the dock and onto the ice, we slid it along toward the channel.

"Hey!" I cried as an idea occurred to me, "hey, wait a second, you guys at the bow! Let's think about this. When we get to thin enough ice, this boat's going to go through, and it would be nice if we didn't. Let's all push from the stern. When the bow goes through, Richie and Mike jump in. Mark and I will hold the lines till we find out whether the motor's going to work." It was a good plan. It worked. After three pulls on the cord of the old fifteen, it coughed its surprise and disbelief at having its winter sleep disturbed. With the next pull it woke up altogether and roared. The boys greeted it with a cheer as Mark and Pat jumped in. All the islands being icebound, they had actually nowhere to go, but that didn't bother them. Lords of the empty river, they buzzed up the open channel to the thin sheet covering the Forty Acres and, returning, nosed their way in as far as they could at other islands. "Just checking ice conditions," as Richie explained in his professional meteorologist's voice.

"We did say once," wheezed Elisabeth, "when we bought the old place, that we could give our children memories. Well, they'll remember this."

Weather always matters up there, but it mattered particularly then. A good storm, a heavy freeze, either one, but especially both together, would make our departure more than difficult. As we collected our gear on the last night, the temperature was clearly dropping.

"And what will we do if it freezes . . . I mean really freezes?" Elisabeth kept asking, and I kept answering with less and less conviction that we would manage.

"Lot more ice," Mark announced as he stomped into the dining room for breakfast next morning. "Wind's coming up too." We had to move, and quickly, or we might really be socked in. No time for two trips, and we couldn't all fit in the punts with luggage. We would have to use the fiberglass and store it with Cuttle till our return late in May.

With customary efficiency, Elisabeth organized the packing and cleaning up. Mark, Richie, and I carried the baggage to the dock, and with the sled Pat whistled load after load across the bay and down to the old dock where Mike stowed them in the punts. The last load consisted of James and Miranda, who whizzed off with shouts of encouragement to their obedient husky. Mark, Elisabeth, and I locked up and packed the fiberglass. As we put in the last duffel bag, we could hear the drone of motors in the distance as the others set off. The wind was stronger now, strong enough to break free big sections of new ice and send them downstream.

"We'd better hurry-hurry-hurry!" Mark advised, "before those floes close up that little channel." It was indeed a narrow passage we had to negotiate. It resembled the waist of an hourglass—open water on either end—but we would have to get through that slot before crossing the last three quarters of a mile to Cuttle's marina.

"We are," his mother snapped, "can't you tell? Bob, did you take care of the electricity?—throw the switch or whatever?" I had.

Everything set at last, we backed out of the notch in the ice that had been our berth, then headed for the slit. Ice growled and grated along our sides as we passed through what a short time ago had been open water.

"Bob!" Elisabeth exclaimed from the bow, "there's a great huge one coming. We'd better get there before it does!"

"Oh, we can beat that one," Mark shouted back. "I'm just wondering how many others got there already. Can you see that passage, Mom?"

After a tense pause, "No. I can't see any passage at all!"

"How about the other boats?" I called. "Can you see them?"

"No sign of them," she threw back over the sound of the engine. "They must have got through and gone around the corner of Foresyth. Can't see them anywhere."

The motor singing at full speed, we scraped and grated along the edges of small floes, and jolted, thumped, and bounced into heavier ones that cracked and growled along the gunwales.

What a few days earlier had grieved me to the point of fury was the Customs official's asssumption that, being blind, I was obviously ignorant, incompetent, in fact, utterly useless. What enraged me now was the sense that he had been quite right. My son at the helm and in skillful command, my wife in the bow with the butt of an oar fending us off and around minor icebergs—and I, I sat amidships, my value consisting in my ability to rock the boat hard and to bounce heavily, thus helping to smash the heavy sheet of ice under the keel. Furious at my impotence, I raged in silence, the knife of my incapacity turning slowly in my vitals.

"That channel was around here somewhere!" cried Mark in frustration, "but it all looks the same." He was standing to have a better look, after having cut the motor down to idling speed.

"Maybe it was," Elisabeth grunted as she slammed the oar butt down on a particularly heavy slab, "but it's not . . . here . . . now!" Fifteen, maybe twenty minutes ago, it must have been open because the others had made it, but it was clogged now.

Kicked up by wind, the waves had broken off fresh sections, floes, whole fields of ice that wind, waves, and current combined to drive downstream to block our narrow escape route. Once closed, what had been a passage became a solid barrier, a table onto which the wind and waves hoisted the edges of new sheets, chunks, and floes—hoisted them up and shoved them crackling, hissing, and splintering along, every moment building fresh mounds and filling in between the mounds to make higher tables, plateaus of splintered glass. And more ice was coming.

"That big floe's almost here!" Elisabeth cried, looking up. "My God, my God! Must be half a mile across! Looks like the whole river's coming! We're going to be pinned in here, Mark, if we don't do something! Don't know quite what, though." A heavy mass thumped against our left side and groaned slowly along it. Mark spun the throttle to full speed, and the motor whined.

"Hey!" I shouted, "slow her down! If the prop hits a big one, we could break a sheer pin!" If that happened, we would be without power, utterly at the mercy of the heavy, grinding forces about us. As he eased back, I felt the ice, driven by its own momentum, nuzzling and thumping beneath the bottom of the boat. On our right, a solid field of rubble, and, like the jaw of an enormous pliers, a whole ice field had come down upon us from the left. As the jaws slowly shut, gradually we began to rise.

"Here we go!" Elisabeth cried. "We're being squeezed right up out of the water!"

"Try to keep the motor running, Mark, and come on! Bounce! Bounce like you never . . . bounced . . . be-*fore!*" Up and down . . . up and down . . . hard! . . . hard . . . *hard!* hoping to keep enough water beneath us for the prop to spin through the crunching, growling, splintering groan; Elisabeth hammering over the bow with her oar butt; Mark giving short bursts of power with the motor to drive the bow up onto fresh ice that, bouncing and rocking, we smashed through and over to a new spot. Foot by foot we crashed, drove, and bounced toward the open water beyond. Five long, boat-buckling, keel-cracking minutes. Then Elisabeth shouted, "We're through! My God! We're actually *out!*"

Twenty-four hours later, in a clean shirt, tie, and sports coat, I sat quietly at a seminar table around which ten students were clustered.

"The only test . . . the only real test of a piece of literature," said one of the students, "is whether it makes the reader undergo the actual experience. If the writing is good, it actually can reproduce reality."

A typewriter chattered somewhere down the hall in the silence that followed. Footsteps clopped back and forth, sounds of a muffled conversation. The radiator behind me hissed thoughtfully to itself over the weary drone of the fluorescent light. The air seemed suddenly too hot, stuffy, dull, opaque. Slowly I got up, took off my jacket, and hung it on the back of my chair.

"No," I said a little sadly as I sat down again quietly, "however good the writing is, it can't do that . . . not *that* . . . not actually."

16

Approaching Autumn

We went up to the island as usual after our schools had closed in 1971, but this time we meant to stay until the weather drove us back to the mainland. I had a year's sabbatical and Elisabeth was on leave from her post for the fall semester. There were several improvements that needed to be made in the Main House to make our prolonged stay more comfortable.

One afternoon early in June, I decided we had talked long enough about taking out the wall between the little dining room and huge pantry to create a sitting room; so I had gone out to the tool bench on the back porch, collected some weapons, and was attacking the wall when Elisabeth came in.

"My God!" she exclaimed, "what are you doing?"

"We've talked about it long enough," I answered.

"But . . . but . . . hadn't you better wait? . . . wait for Durrell or somebody? I mean, isn't that . . . didn't you decide that was a bearing wall or something? Couldn't the roof come down if you knock out that wall?"

"It could," I admitted, splintering the first ten-foot length of knotty pine. She gasped.

"Well, then, hadn't you better be careful?" she urged.

"I am being careful," I countered. "This tongue-and-groove paneling isn't easy. They nail through the tongue, you see . . ." I began to explain.

"But the roof! What about the roof?"

"It'll be all right until I take out the studs, and I won't do that. I'll just put in a jack . . . some sort of temporary support until I get in a new beam the whole length of the wall. I'll have to put in posts or something at each end for the beam to rest on, and then I'll take out the jack. See? Everything'll be fine."

"I don't think you know what you're doing," she said firmly. "I mean, it's all very well to experiment, but this is our house."

Her obvious confidence in me was most reassuring and, of course, entirely justified. Thus buoyed, I pressed ahead, she wincing at every groan and crack and squeal as nails and wood parted company. I pressed Mark into service. Together we hammered, pried, and wrenched. To James fell the job of pulling any nails we had missed and "carefully, I said *carefully and neatly*" stacking the boards on the back porch. We did our best not to damage the boards because, recut, they would be used to close in the entrance from the kitchen into the pantry.

By the time the paneling was stripped from the studs and the old entrance to the pantry closed off, Elisabeth's excitement had overcome her fears. She walked back and forth between the studs, praising the roominess of our new arrangements, debating where to put the "new sofa," and what color to paint (maybe painting wasn't such a good idea . . . maybe just staining) the new paneling on the inside of the new room because, after all, we were going to panel it, weren't we? As her imagination took fire, the heat of my zeal diminished proportionately: the really ticklish part—the removal of the supporting studs and the insertion of the new beam—was yet to be done.

"I don't know what you're so worried about," she said. "It's going to be lovely."

"Well," I hedged, "it's the roof, that's all. These things have to be thought out, you know." I did think them out—or, at least, about them, if not out. Long and hard, I thought, but no amount of thinking, of measuring and remeasuring seemed to put a single sliver of wood in place. We even went so far as to pick up the beam and the two supports from Mitchell & Wilson in town, bring them over and put them on the back porch, but even that didn't seem to get the job done. Finally, on a Saturday afternoon, driven more by shame at my own cowardice than by courage and confidence, I dug out the bumper jack from under a pile of old lumber and knocked together a T support, which I nailed in place. Under the leg of the T, I put the jack and levered it up until the ceiling creaked.

"Okay," I said, loud and clear, "okay!"

"Okay what?" inquired Elisabeth.

"Well," I said, ". . . I'm ready to cut out these studs . . . just about ready. . . . But maybe we ought to have a little supper before we start, because once we start, we're not going to be able to stop until . . . until it's . . . until it's done." She agreed, and I took a leisurely pause to munch and to think one last time. But that couldn't go on forever, so finally I swung irresolutely into action with the saw. Keeping as close to the floor as possible, I worked through the first stud. To my chagrin, the saw came out the other side, but pressure from the ceiling simply pushed the upper part of the stud down hard upon the stump.

"What do you do now?" inquired Elisabeth.

"This!" I declared bravely, and gave the long upper section a sound whack with the hammer. Not sound enough, though. It took two more whacks before the stud hung loose, like a reluctant pendulum. Still more bravely, I seized it and worked it free from the spikes that held it in position at the top.

"There!" I cried triumphantly, waving the eight-foot hunk of two-by-four, "and the roof's still up!" It was, too, but this was only the first stud. There were seven more to go. I tapped the leg of my T support. It was rigid, as if nailed to the jack by pressure from above.

"Does this look straight?" I inquired earnestly of Elisabeth, patting the leg again.

"I don't know. Yeah, I guess so. Anyway, how am I supposed to know?"

"Well, you can look at it, can't you? Because if it isn't . . . if it isn't, the pressure will kick this jack right out from underneath, and down she'll come."

"Down what will come?" she asked nervously.

"The whole business . . . I don't know, but then we'll really be in trouble."

"Maybe you'd better wait till . . ."

"Just keep back," I said, taking the saw to the next stud. "Keep back but keep watching, and let me know if this leg starts to bend or anything."

"Hey, what's going on here?" interrupted a voice from the door.

"Oh, Ben!" cried Elisabeth with relief, "Ben, he's, he's at it again . . . he's going to bring down the roof or something."

"Holy Jesus!" was Ben's thoughtful comment to himself as he surveyed the destruction.

Dropping the saw, I rose from my haunches. "There you are, Benjy! Cast your professional eye over a professional job."

"Yeah," was his abstracted reply.

"Tell him . . . you tell him, Ben, what he's doing," she urged faintly.

"I just . . . damn it, Russell, I just came up for a quiet beer, and here you are tearing up the pea patch. Let's have a look at your support there." As he examined it, I went to the fridge and opened two beers. "Here," I said, thrusting the beer toward him.

"Don't drink while I'm working," he replied, "and neither should you. But this looks all right—pretty good, in fact."

"You mean he's done it right?" Elisabeth asked incredulously. "You mean the roof won't fall in?"

"Might," he admitted, looking up. "Could, I suppose. Have you got another jack, Bob? Wouldn't hurt to put in another leg under your temporary plate." I scurried to the back porch, fished about, and came up with an old one of Jack Stuart's that I had borrowed the previous year and that Jack had been after me for. As we put it in place and cut a two-by-four to the right length for another leg, I explained about the jack.

"Good for you," he congratulated; " 'Return only when forced.' That's my motto, so be careful about lending me anything. Here, just another twist on this bar." The ceiling complained. "Good. Back at it with the saw." Another stud . . . another . . . another, and as they came out, the supporting legs sang with tension.

"But Ben," put in Elisabeth, "you don't have to do this. Sit down and drink your beer. He got himself into this, and I bet you've been working all day."

"Yeah," he admitted, "but I'd kind of like to see what's going to happen myself. It's not *my* roof, you understand," he chuckled, "but I'd still sort of like to know. Go ahead, Bob, and I'll get the uprights and the beam. Wasn't that the big stuff I saw on the back porch?"

"I'll give you a hand," said Mark, and he and Ben dragged in the lumber while I buzzed through the last stud.

"Don't knock the last one out till me and Mark put the uprights in place ready to take the beam because . . . because . . . the sooner that beam's in, the better I'll feel."

"Why not put in the beam before, actually before you take out the last stud?" Elisabeth inquired. "I kept telling him to do that, Ben, but he wouldn't."

"Can't," Ben laughed, "the studs are in the way. The beam's got to go where those studs were."

"Oh," she said, unbelieving, with a you-too sort of tone. Ben and Mark tacked the uprights in position with small nails, and Ben climbed up on the stepladder with one end of the beam at the opposite side from where I was ready to pull the last stud. He tried to slide the end between the top of the upright and the plate to which the studs had been nailed.

"Jesus," he groaned, "what a fit. These damn people who measure everything down to the last sixteenth! Can you hand me that hammer, Liz?" and he whanged away, driving the beam into position. Meanwhile Mark was standing on a chair with the other end in his arms. "Okay, Mark. You get your end up there and be ready to slide . . . I mean slam her into her spot. You weren't making a cabinet or anything, Russell, you know. It didn't have to fit this tight! Better give Mark the hammer, Liz. Sorry, I've got to stand here. Okay, Bob, pull her out." With a mighty yank, I pulled out the last support and began working it free from the nails at the top. Mark jammed his end of the beam into where it was supposed to go, but it wouldn't.

"Ceiling's starting to bend, ceiling's starting to bend," Elisabeth warned.

"Yeah, geez, I see it," Ben cried. "We need another jack—that's what we need—so we can get that bend out of there. We'll never get this beam in unless we can get that bend out. Danny! Take the boat and buzz down to our place and bring back our jack. It's under the front porch. On the double. Fly! We don't want Mr. Russell's house to fall down . . . leastways not while we're inside, so *hurry.*"

"Now what do we do?" asked Elisabeth.

"Might as well drink my beer," Ben said. "We can't do anything till he gets back, and the ceiling doesn't seem to be bending any more."

"Even if it does," I put in, "we might as well drink a beer for all we could do about it right now."

But it didn't cave in before Danny got back, and soon we had a third support and, with it placed in the middle of the bend, we jacked till it was straight. Even then, Mark's end of the beam wouldn't go in without the persuasion of a small sledgehammer, but it did go in at last and our arch was complete. As we slowly released the jacks, the new beam groaned.

"What's that noise?" Elisabeth asked nervously.

"It's just accepted responsibility," I answered.

"Will it hold?" she pursued.

"We put it there, so it better had."

"Huh," was her only reponse, and she began sweeping up the rubble of bent nails and splinters.

"Some room," Ben admired. "Who'd have believed it? Only thing is," he added with relish after a pause, "what are you going to do about the floor now?" I said I wasn't going to do anything about the floor, but he hadn't been talking to me. "Can't paint it! Not again." Elisabeth agreed. "Geez, that camp must have got a great buy somewhere on dirty gray paint, the way they sloshed it around."

"Christ sake! Can't you shut up, Ben! I haven't even got the stumps of these studs out of the floor and you're pushing for a refinishing job!"

"Right," he said cheerfully, "and I'm leaving for Toronto in the morning, but it would be nice, wouldn't it, Liz? I mean, now you've got such a lovely big room—great sitting room, really. But that floor!"

He went back to Toronto the next day; by that night I was scraping at the old paint.

"Beautiful!" he congratulated two weeks later. "Didn't I tell you it'd be beautiful? Not hardwood, but she does look lovely. I'm proud of you, Liz. Of course, there's the kitchen floor. . . ."

"'Go to hell!" we chorused.

He laughed. "But you got to admit there's nothing like getting down to bare wood. You're going to panel, aren't you, Bob? In here, I mean . . . in the sitting room? No sense in leaving a job unfinished, I always say."

"Put down my beer and get back to hammering on the roof of your shack," I threatened.

"No, but seriously," he went on, "that reminds me. Remember the time I explained to you how to shingle? You made me cut up those bits of paper and actually show you? Remember? Well, I hope you do because I don't. Honest to God, I don't, and I got to put shingles on the west wing I just finished."

"All I can remember is that you said something about beginning at an edge and cutting the first shingle."

"The first! Why would I have to cut the first one?" he asked.

"How am I supposed to know!" I complained. "I asked you the same question five years ago and you couldn't give me a straight answer then. You're supposed to be the expert!"

He reflected for a moment. "Augh, I'll ask somebody, but you're going to have to fill in those cracks, you know."

"Cracks? What cracks?" I asked.

"Maybe you can't find them now, but they'll find you, come October. Boy! the wind will come whistling up through that floor! You know, Liz, I always used to wonder why the women up here sat so funny . . . you noticed them? They sit with their feet all tucked up under them. Yeah, well, I've found out why. It's because the floor gets so goddamn cold! That's why! The footstools! That's what footstools were all about in the old days before insulation. Laugh! Go ahead, but you wait! You'll find out in October that I'm right. You're going to need some heat around here . . . more than that little wood stove. That's enough maybe to keep the chill off this room, but how about the big kitchen? What you ought to have is one of these oil space heaters out there."

"He's got one of those!" said Elisabeth. "He's had it for two years. Just hasn't installed it, that's all."

"Then you're away, Bob," he congratulated. "Put her in, I would. . . . Did I start something?" he asked with amusement, sensing her anger. "Anyway, no kidding, I'd put it in, I would. But remember, I'm not involved. I didn't have any part in whatever it was. Still," he chuckled, "I'd put it in. Don't hit me . . . don't hit me, Liz! but I really would. Good night," and he left.

Elisabeth's accusation that *"he* has one," dated back two years when she had noticed a space heater advertised in the *Reporter,* Gananoque's weekly paper. It hadn't said whether it burned oil or wood. If it had admitted to oil, we never would have gone to look at it. She didn't like oil, didn't like the way it burned without any crackle or snap. It wasn't friendly, comfortable, had no personality, wasn't human like a proper fire. I conceded all that, but countered by pointing out that those things really did give off good, steady heat and there was no kindling, no chopping, and no ashes. At this, she retreated to visual grounds, declaring they looked awful, however much heat they gave out, and she would have to look at it. And besides, they blew up! just plain exploded and burned the house down, and had I ever heard of a wood or coal stove doing that? Had I?

So long as the discussion remained *in abstracto,* it offered reasonable grounds for amusement, but when we arrived at the house to find that the heater advertised was in fact a burner of oil, a new dimension was added.

"You know what I feel about them," was all she would say, too low to be overheard. I did know, but this one, this particular one, came equipped, wonder of wonders, with a real live thermostat plus a fan, and they said the thermostat would turn on the fan, heat up the room, then shut off till the room cooled down a bit, and then it would, all by itself, turn the fan on, open the fuel valve to make more fire, and heat up the room again without my having to do anything. All that sounded so incredible, and the seller seemed so convinced it really could and actually would do all those things—plus the fact that it was only thirty dollars—that I bought it and we carted it back, Elisabeth maintaining a stony silence all the way. A little sheepish—it would be she, after all, who would have to look at it—I had stored it in a little junk room in the boathouse where for two years I had kicked and stumbled over it whenever I went in for my fishing rod or a raincoat.

But now, now was the moment, the moment for actual installation. Had we not, I thought to myself—had not Durrell and Norman, with me as helper—put in the big picture window in the kitchen wall for her to look through at breakfast, lunch, and dinner, in fact, all day long if she wanted to? Her eyes need never stray slightly to the right, to be offended by the quiet, dutiful, and very warming space heater with its thermostat! A little softened by Ben's urging, perhaps even by the force of argument, but also by the lapse of two years since its actual purchase, she agreed, with the proviso that I fully understood her opposition and that I must never expect her either to like or admire it. Agreeing to the conditions, I approached Norman, our resident holer. He was half, but only half, preoccupied with tying up the details of a trip to Rome. He had nailed down a fat grant from the Canadian government to spend a year doing research in the Vatican Library on the papacy, and was arranging housing, schools for his two children, Jeff and Mary, plus relearning Italian every morning on the porch of the Infirmary with Betty, his wife. The previous winter he had written to announce the grant and gloat a bit, and I had actually answered and extracted a solemn promise that he wouldn't waste all his time on the papacy. I pointed out that Italians were great masons and bricklayers and that unless he learned more than the mere rudiments of this art, he could spend the next summer after his return howling on the end of the Bay Street Dock in town. He replied, acknowledging the solid value of some one of us knowing masonry, confessed he had always wanted to know it, and that, indeed, acquiring this skill had been half his reason for submitting his proposal for the grant to go to Rome. Meanwhile, till he had mastered the art of slinging cement, he would stick to hammer and saw, even if it was *my* saw. It was, too. I had given myself a good saber saw for Christmas, and with it Norman had already cut a lovely hole, almost a perfect circle, in the ceiling of the freshly completed sitting room for the stovepipe from the little wood- and coal-burning heater I installed there. Now that the space heater was to go in the kitchen, I needed a hole in that ceiling too.

"Sure. Let's go," he said, putting aside his oboe with which he had been disturbing everyone within earshot. (He played first oboe for one of the Toronto symphony orchestras.)

I climbed up into the floorless attic and crept along the joists until I reached the spot beneath which he was banging with the end of a stick.

"Somewhere around there," he called up. Since it wouldn't be such a good idea to cut through a supporting joist, and since he couldn't tell where they were, I measured the distance between the right ones and drove a nail down from above to mark the center of the circle. He drew the circle with a pencil (which reminded me how jealous I was of anyone who could mark a piece of wood so quickly and easily with a line they could see to follow), flicked the switch on the saw, and the blade chattered dutifully along his line. He moaned and groaned about how all the sawdust was falling in his face and none of it on mine, but the hole was soon cut, and it was a beautiful fit.

"Yeah," he said, looking up to admire his skill, "but where does the pipe go after that? Those fumes can't just go into the attic!"

"You're a great holecutter," I congratulated, having climbed down, "but I've thought the job through just like Durrell does before he picks up a tool. The pipe from the stove in the sitting room goes up through the ceiling through your first hole, then right along and into the chimney. I've got a T section of stovepipe I'm going to put in that line up there, and connect the pipe from the space heater into the third leg of the T, so both heaters will exhaust right into the chimney."

"By God!" he cried with admiration, "you're not a hopeless case after all, even if you are only an English professor."

So that's what I did. (That simple sentence glosses over hours of teetering around on joists while sweating, struggling with, and swearing at sections of pipe that were too long or too short, that wouldn't go together at all, or else were stuck so tight they wouldn't come apart; but anyway, it did get done.) Finally, smug, but still a little embarrassed under Elisabeth's eye, the space heater with its thermostat crouched beside the picture window waiting to be called on for heat. Though the thermometer declared the temperature to be seventy-five, we had to light it to prove what it could do, and it did. What's more, it did it all by itself, just like its previous owner had said it would. It brought beads of perspiration to our happy faces—well, Norman's and mine. Elisabeth merely observed she hoped it would do that when it got really cold.

Late in August, the evening before the departure of Norman and his family, Durrell and Arvilla invited us all down for a good-by drink. Amid the cheerful and brisk intellectual discussions about the proper mixtures necessary for building a chimney, inquiries about the precise characteristics of seal-bond, and wonderings about whether the Romans ever solved the problem that sporadically fixed our thoughts—pouring cement under water—the telephone rang. Durrell took it.

"Yeah. . . . All right, hang on. We'll be right up!" As he hung up, he was headed for the door. "That was James. He came in and found the house full of smoke!" A scrape and clatter of chairs, then the rumble of many feet as the stampede developed. Across the flat ground past Durrell's dock, up the narrow, rocky path to the high ground, along behind the Infirmary, and down the slope to the Main House we poured. Durrell was already coughing inside when we arrived, smoke pouring through open doors and windows.

"It's all right!" he called. "Can't see any fire." I grabbed the stepladder, set it in place, climbed up, pushed open the trap door, and hoisted myself into the attic, where I sniffed my way along the joists to the stove pipes. The one from the wood burner was warm, but then it ought to be, because I had left a slow fire in that stove to take the chill off the sitting room. There certainly was smoke up there too, but that could be coming up from the kitchen. The odd thing—yes, very odd—was that the pipe from the space heater was warm. It had no business to be. That heater wasn't lit.

"Here's the trouble," I heard Durrell saying from below. "The smoke's pouring out of this space heater. It's not lit, but there sure as hell is smoke."

"There's a little fire in this one," came Norman's voice from the sitting room.

"Robert Russell!" called Durrell in his most formal, professorial voice, "I have a question to ask you when you get down here."

"I'm not coming down," I answered, "because the answer is yes. On second thought, I will come down because I clearly remember asking your opinion, and you said you thought it would be all right to hook up the pipe from the space heater to the one from the wood stove."

He stamped his feet. "I said it might be, and it might have . . . only it isn't." And, of course, it wasn't. Most of the time it wasn't too bad, though. It was just when the wind came from a certain direction, which, for some reason, created a downdraft in the chimney. When that happened, you could smell kerosene in the sitting room and wood smoke in the kitchen. That wasn't quite the way Elisabeth put it.

"Drive us out!" she groaned and coughed. "Your father's trying to smoke us out like rats, Miranda."

"You don't smoke rats out," I objected, "and anyway, it isn't that bad!"

"It's pretty bad," Miranda would say, sadly opening another door.

"You put in these damn stoves," Elisabeth said—by which she meant the space heater—"for warmth, and then they smoke so bad we've got to open all the doors and windows! What's the point of that, I'd like to know."

This wasn't exactly a constant barrage because, luckily, the wind came from the wrong direction only every few days. Then she would make her point again, always closing with, "What I don't understand is, if you know what the trouble is, why don't you *do* something about it!"

"It takes time," I kept saying a little weakly, "to do something about *everything*."

17

Miranda and Prospero

"All we can ever really give our children," I had urged to Elisabeth back in '56, "is some happy memories." Sentimental? Yes, but also, I thought and still think, true. More important, it helped to persuade me that, while neither economically sensible nor perhaps even feasible, we nonetheless ought to buy a piece of the Thousand Islands. It worked. We did. And *we* worked. Quite easily, we unconsciously settled into a pattern—committing ourselves furiously to our jobs and all they entailed from September through May, but ever and always, underneath the turbulence of our surface lives, was the peaceful certainty that June would come and that, when it did, we would grievously overload the car, at the very last minute wrench, struggle, and fight out for ourselves places to sit, and then take off. Hours later, cramped, exhausted, fed up with the dog and the cats that had refused to sit still, we would finally roll to a stop in the parking lot at the Bay Street Dock and fling open the doors to gulp the first sweet lungful of St. Lawrence air. We spilled out, joyous in the knowledge that three whole months of the island lay before us, that they belonged to all of us as a family. This was our place, our time, our world.

"Trouble is," said an older islander with sadness, "they grow up, and when they grow up, they leave."

"Catch me!" Richie said with a lazy laugh. "But you know, I

never realized before [he was a freshman in college] about April, everybody on campus starts moaning and groaning about summer. All those kids hate summer. They do! I couldn't believe it. And they couldn't believe it when I told them I just waited through winter for the summer to come because summer is all that really matters!" He meant it, and so it was good to hear. But for the last two summers he hasn't been with us. What's pulling him away is his life and he must follow where it leads because it's his. For him that means, at least for a time, being elsewhere.

Mark, at least in one sense, left even earlier, but he didn't leave the islands, only us. At fourteen he spent his first summer as boat-boy, caretaker, and companion for an elderly lady on Kalaria, a two-acre island just off the southwestern tip of Hay. Mohawk includes that tip plus even a tiny island in the narrow channel separating Hay from Kalaria; so, while Mark is making his summer memories only eighty feet from our property line, nevertheless he is no longer a regular member of our Mohawk family.

My mind peels back more than a decade to one blazing afternoon in Lancaster when, for some reason, I had to return to the college. Mark put his small fat hand in mine to come as far as the first corner, crossing the street being still forbidden. Silently he trudged beside me under the shade trees in front of the row of neat brick semidetached houses. Through a screen door drifted the tense uncertainties of dialogue from a TV soap opera to mingle with the lazy cadence of a handmower in someone's back yard; the soft, intermittent exchanges of two small girls playing with their dolls on a quiet porch; the distant, occasional whoosh of a car; the smell of freshly cut lawn, and the long, shimmering blades of sound with which the cicadas mechanically announced the peace and heat, the quiet and loneliness of a hot afternoon. In my hand, the hand of my son. Beside my heavy footsteps, his light tread. Unspeaking, we waded through the afternoon that stretched as far as the corner where he paused—he, forbidden to go farther; I, required to go on. Required? Required by what or by whom? Not clear—not at all clear, and yet required.

"Well," I said, "I'll leave you here. Thanks for coming." I bent and kissed him, and he unabashedly returned my kiss with a large, wet mouth.

" 'Bye, Dad," he said, but did not turn to run back through the emptiness of afternoon to his toys or his mother or his friends. Small, thoughtful, silent, he stood as if to watch me safely on my way. Reluctantly submitting to the pressure of vague, but heavy necessity, I stepped from the curb and began crossing the street. Alone, a child, my son! and I was leaving him solitary on that corner of early summer. He coughed. I turned and came back. I could not bear to leave him standing there. My vague but deep distress came, perhaps, from an unconscious recognition that our roles were reversed. It was not I who should be leaving him, but he who should and must leave me—a fierce, brief preview of a basic parental grief. The principle of growth would require it, but he was as yet too young. I must instruct, must even order him.

"Are you all right?" I asked, bending down.

"Yep," he answered.

"Then you run along back, eh?" As he turned and walked slowly away, I waited, wondering if this would be in later years one of those moments that suddenly and for no reason flashes upon the screen of memory, a still picture, a vignette, a window through which we may see into . . . into what? If it was to be one of those moments for him, what would it mean to him? Anything? Nothing?

Aimlessly, I wandered on my way to do and be whatever necessity required, but wondering.

Now, more than ten years later: "Here comes Mark!" cries Miranda happily over our slightly solemn breakfast. Irrepressibly joyful even on Thursday mornings—the time divinely appointed for housecleaning on Kalaria by more than seventy years of habit—he sings or shouts his way up from the dock to borrow the vacuum cleaner. Not recognizable songs or even words, but cries of sheer animal delight—barks, howls, yodels, as he swings along the path between the pines and hemlocks toward the Main House to sit through one, two, perhaps three cups of tea amid the litter of our breakfast while he recounts with incredulity fresh enormities of his employer's Victorian courtesy and frugality.

"You've got to remember, Mark," says his mother, pouring him another cup, "that she's done things that way for forty years, and . . ."

"*Forty!*" he explodes. "Seventy-five's more like it."

"Yes, but people are funny in lots of different ways," says his mother, "and you've got to learn to be tolerant. You've got a pretty soft spot, you know. How would you like to be working three thirty to twelve in that can factory with Richie?"

"No-o-o thanks! Not me. Bet Lancaster's hot. Have you heard anything from Rich?"

"Nothing except that he's thinking of quitting and hitching to California."

"He can have," Mark says, leaning back with a sigh, "California and everything." But I wondered about the next summer or the one after that, and then the string of summers—how many?—before he would find his way back to the rocks and pines of Mohawk.

James, thirteen, would still have a few summers with us. He and Danny were inseparable. When Ben had finished his cabin, James had been their first guest.

"And they woke me up the next morning, too," Ben complained. "Know what they were doing? Throwing an empty Coke can up onto the roof and taking turns catching it. I mean, how can a guy sleep with that kind of noise?"

"Did you tell them off?" I asked.

He laughed. "Gave 'em hell, I did. Yeah, sure I did. Beautiful! that's what I said. That's what it's all about, isn't it? I can remember when I was their age. Beautiful. I'll shingle roofs as long as they throw cans. It's when they get too old to throw cans, that's when I'm going to start worrying. Maybe then Pat'll come down again. Bob, she's got to come because Danny's going to grow up, and then what will I do?"

"You could always add another wing or another porch," I suggested.

"What for, if they're gone?" I had no answer. "And what are you doing all this for yourself, if it comes to that?"

"Retirement," I answered. "I'm going to retire someday and live up here."

He thought about that.

"If I believed you—which I don't—but if I did, maybe that's what we ought to plan for, an old folks' home. Kids have their communes nowadays. Why can't we have ours? Why, you've practically got one now, with Norm up there in the Infirmary and Durrell down at the bay. There doesn't *have* to be kids," he paused, "or does there?"

"Anyway," I said after a pause, "the time is coming when there won't be."

"We're not going to have to take up bridge, are we?" he groaned, "not *bridge!* Before I'd do that, I'd burn my place down and start again."

"Danny and James are still here," I encouraged, "and there's Miranda and Emma, [Emma was Danny's eight-year-old sister] so we've got a while yet."

"Yeah, but we ought to plan," he mused. Then, rising, "I'm going home to drink a Coke. Send James down first thing in the morning. I'll leave the empty can on the porch," and he left.

He was right. We ought to plan. But we had to see what it would be like without children before we could plan. I arranged for a sabbatical from June of '71 until September of '72. Since we couldn't live on my half salary, Elisabeth got a leave of absence for only the fall semester, and we set about testing our capacities for island living minus the regular stream of guests and the summer buzz of the much-touristed river.

As for children, Richie would be in college at Franklin and Marshall. Mark, very much at his own request, would be off at a private school; James, since he had won a varsity letter in soccer in the seventh grade and was likely to win one in eighth and ninth, would be, if permitted to go to the same junior high, a prime candidate for a soccer trophy in two years—an award never yet given at that school. He pleaded to be allowed to go back: since Danny wasn't going to be on the island and since he would be the first in the history of the school to get the trophy, could he please go back to Lancaster in the fall? Hearing rumors of this, Durrell and Arvilla Enck offered to take care of him. Miranda, at first joyful at the thought of spending the fall at Mohawk, began to wonder. If James was not to be there, she would have to face all the uncertainties of a new school alone. Then, too, upon our return, she would have another new school, the junior high, to face by herself. Other Lancaster friends offered to keep her until our return but the prospect of being separated from both of them made us rethink our scheme. Without the heart to deny James his chance at the trophy, and unwilling to force Miranda to stay, our whole plan of spending autumn on the island was teetering.

One afternoon late in August, Miranda and I were alone. The fall exodus of islanders had begun and those who had not yet left were packing and closing up.

"How about a walk?" I suggested. A little surprised, she agreed, and we set off toward the point beyond Richard's cabin with Ginger, the dog, in tow. The portliness of middle age had done nothing to dampen the delicacy and gaiety of the poodle that lurked somewhere in her nearby ancestry. Ginger had long since made it her business to let none of her family—us—escape through a door without her somewhat shy, but resolute attendance—shy, unless openly invited to come, which in this case she was. Responsive to our every mood, she trotted along discreetly a few paces behind us, as if understanding that this was a serious occasion and, while not presuming to intrude, she wanted to stay close enough to participate if the atmosphere should change to one of fun and games.

A good breeze out of the southwest brought an occasional whitecap in to break against the low cliffs to our right and made rich commotion in the hemlocks and pines lining the shore. Though not cold, the air carried a promise of cold, but in the patches of sunlight screened from the wind were pockets of summer, warm and full. Through open arcades, through tangles of juniper ferns, we moved from season to season as if they, like the girl beside me, were balanced on the ridge of present uncertainty. We both knew which way they must go, and, I, older, knew the way she, too, must go—but only eventually. Was "eventually" now? Vaguely, perhaps, she sensed that the beginning of school in Lancaster meant that the stream of her friends, after its summer pause, would be on the move again. She belonged to that stream; her destiny lay with it and in that country of the future toward which it flowed. It flowed from the sun-drenched peace of childhood toward the drama-rich uncertainties and promiseful excitements of adolescence. That she would go, must go, had to go, was clear, perhaps, even to her; but the imperative call of growth is also frightening. Part of her wanted, needed to stay, to hold back.

Sensing this turbulence in her, and not sure whether a father's rights included the prerogative to forcefully interfere—knowing, too, that I had only at my disposal the arguments that could be cast in words, and having not much faith either in arguments or words—I had suggested the walk in the hope that being out together in the afternoon would do what speech could not.

To walk with her was to be taught again the loveliness of the earth and to re-experience the excitement of being young: alert, keen, eager for delight, and, with electric-quick responses, finding it everywhere—the soul straining up and out toward the world so hard that the skin can barely keep it in place. The spring of matted pine needles under the bare foot . . .

"Ouch!" she cries, hopping, but only for a moment, to rub her foot. "Stubbed my toe!"

"But why don't you wear shoes?" from the sluggish majesty of my solemn age.

"Don't like shoes. Come on. It's all right." And we go on, I knowing why she doesn't like them, suddenly uncomfortable in my own, angry at the wisdom that taught me wearing shoes prevents the stubbing of vulnerable toes, but lacking the courage to take mine off. Stepping over the trunk of a cedar fallen across the path.

"Feel, Dad! Feel. Just feel how smooth it is. Like somebody sanded it all down. But cedars are like that. Wonder why cedars are like that . . . so smooth. Here's some prickles. Is this juniper? Smell. Doesn't it have a funny sharp sort of smell? Wait now. We've got to go under this one. Duck. Blown over, but it's propped up by that other tree, and it still has leaves. Wow! Blown down . . . well, not down, but still growing, still having leaves."

So she leads me, carefully and with more kindness and patience than one has a right to expect from youth, over and around the windfalls of past autumns (a journeying forward for her, and for me backward) till at length we swing around the last pine tilted back by years of pressure from the southwest winds, out onto the huge, triangular slab of rock whose apex is our point, thrusting itself toward the open water of the Forty Acres. We sit down cross-legged on the tip. Here we take the full force of the breeze and the spray from the waves that come glubbing and chunking into the cracks, crevices, and little caves under and about us in the rock.

"Listen!" she directs; "all the different noises!"

"That's because of the different sizes and shapes of the hollows," I explain.

"I know, but listen how strange!" I listened and heard while she sat silent. A wet nose touched my hand, and Ginger's warm tongue inquired with two tentative licks if it was all right for her openly to join us.

"Oh, Ginger," cried Miranda, hugging her into her lap; "you can sit with us. You'll get wet, but you can sit with us." Then suddenly, "Dad, look here! these circles in the rock!" She took my hand and showed me the two-foot-wide set of concentric circles drawn as if by a sculptor's chisel on the slab. "You think Indians did that?"

"Mm! Probably not. Maybe . . . I don't know . . . maybe a whirlpool or something thousands of years ago when all this was under the sea."

"Wow! Do you think so? Imagine!" After a pause, "Hey, can we go along the channel?"

"Sure," I said, rising, and we went back to the base of the triangle and the trees to follow the shore into the narrow, sheltered slot between Hay and Kalaria. For the first few yards she picked our way among boulders. Soon we were on sand, the sound of the waves against the point far behind us.

"Gosh, the water here is so still I can see my face in it," she exclaimed. "Amazing. And right back there . . ."

"Bet the water's low enough to wade over to the little island," I offered.

"Do you think we could? Oh, but you've got shoes."

"I'll take them off," I said, sitting down.

"Great, but you'll have to roll up your pants," she said as I untied the laces.

"I've done that before." As I made my preparations and Ginger nervously inquired with a wet nose into exactly what this was all about, Miranda stood gazing.

"So quiet here. The rocks so gray, silver, so still."

"Okay," I said, getting up, and together we stepped into the still, clear water, the mixture of sand and mud squidging up between the toes. The island fifty feet away was a picnic spot for Miranda and Norman's daughter Mary. With sandwiches and bathing suits, they had spent long days lazing on its rocks, paddling in the shallow water between it and Hay, and swimming in the deeper water on the other side toward Kalaria. They had named it Emerzym, combining letters of their names to express the tangled privacy of their experiences there. But Mary had gone, and we were alone—alone except for Ginger, who began whining and barking unhappily at the water's edge.

"Ginger wants to come," cried Miranda, half turning to go back for her.

"No, no. If she really wants to, she'll come."

"But she's miserable! Listen to her."

We—rather, I—pressed slowly forward, carefully testing the footing at each step with the caution born of experience. She walked with the assurance and vision of youth. Halfway across, she suddenly stopped and bent down. "Dad!" with delight, "here! Look! Acorns! A little ring of acorns on the bottom—five of them—a perfect circle—just like somebody arranged them there. She picked them up and in a dripping hand presented me with her find. "Who could have put them there?" She gazed about. "No oak trees hanging over the water. They couldn't have just fallen in. There're no oak trees for them to fall from—and so perfectly arranged. Somebody, somebody did that, but who?"

"Very mysterious," I agreed.

"They're beautiful, aren't they? So smooth, and they shine. Let's plant them on Emerzym. Oh, Ginger," in a pretense of complaint over her shoulder; then laughter, "She wades in up to her stomach, then runs back. Come on, silly; swim! Scaredy-cat!" she taunted lovingly. "Come on."

"You've got to remember she's middle-aged for a dog," I said. "Also she's . . . well . . . nervous about things."

"What's there to be nervous about here? There are plenty of things to be scared about, but not here."

"She's afraid of being lonely," I said, "being left over there all by herself."

"But she can swim," Miranda objected, "I know she can. I've seen her. Shall I go get her?"

"No. Middle-aged or not, she's got to get over her fears."

"But will she?" Miranda implored. "I could easily get her."

"She's got to do it herself," I said.

We waded out onto the broad, flat shelf of rock and sat down on a boulder. I felt around in the crevices where tufts of grass declared the presence of some earth in which to plant our acorns.

Digging with my fingers among the roots, I shoved one smooth little barrel shape down and carefully replaced what I had disturbed.

"Do you think it will really grow?" she asked excitedly. "I mean a real, whole, big tree? Give me a couple. I can find some better places." I gave her three. She paused, still bothered by Ginger's whining and barking. "Oh, Dad, I'll have to get her. I can't stand it," and she moved back toward the water.

"Don't carry her," I commanded gently.

"No. No, all right, but I'll go to meet her, encourage her. She just needs encouragement." She strode out, calling to Ginger, pleading with her, praising, reassuring her, and finally the dog pushed the last bit of earth from beneath her feet and swam. Halfway across, she turned and started back for Hay.

"No, no, Ginger! No. You're halfway. You don't know it, but you're halfway; you're more than halfway. Come on. Come on. This way. That's the girl. That's a good girl! Good *girl!* Just keep coming. You're going to make it. You are; you really are. That's it, that's it, that's it! Hurray, *hurray, Ginger!"* and she clapped and danced and shouted as, puffing and sneezing, Ginger dragged her small, tubby body into shallow water, then exploded up the rock to where I sat. She flung her sodden bulk into my lap, wriggling and snorting with happiness and pride in her achievement. *"Yeah!* She did it. Yes, you *did* it, Ginger," cried Miranda, bending down and hugging the soaking bundle of joy. "Where are those acorns? No, Ginger. You go shake somewhere else for a minute. I've got to plant some acorns."

"Not easy to be brave, is it, Ginger?" I said, rubbing the shaggy head she shoved into my hand. . . . "Not at our age."

After a half hour of indolent exploration, we waded back to Hay, Ginger needing no encouragement to follow. Contentedly Miranda sat beside me as I scrunched my feet around in the sand, drying them, then brushing them off, then putting on socks and shoes.

"It's not even that we went there all that much, Mary and me," she said thoughtfully of Emerzym. "Important thing was, is . . . is that it's there to go to when you want to. A nice feeling to know that it's there."

"It is," I agreed.

We spent another hour leisurely picking our way along the

shore around into the back bay—so called because it is on the other side of the island behind the Main House and string of cabins. We passed the tree in which almost every evening we could count on seeing a pair of herons.

"Bet Indians pulled their canoes out here," Miranda said, thumping the wet sand with a bare foot. "I bet they did. I can just imagine them. Wonder what this was all like then? Wish I could have seen it then, maybe two hundred years ago."

"And maybe two hundred years from now there'll be a girl standing here wondering about us," I said, "wondering what we were like, what we did."

She laughed. "I'll tell her, shall I? I'll write it here in the sand with . . . with . . . here's a stick. *'We had a good time.'* There. Okay? Shall we go back now? Mom ought to be coming home soon, and anyway, it's time for a cup of tea."

As the kettle began to boil, Elisabeth arrived with the groceries. We brought them up, put them away, and gathered about the table.

While Elisabeth was pouring second cups, Miranda said: "Oh, Mom. I've decided to stay."

Elisabeth stopped pouring. "Have you, darling? That's wonderful. That's really marvelous. We'll have such times."

18

With Miranda Still

Pow! The tiny cannon shot of a beaver's tail exploded on the water twenty feet to starboard. Miranda broke the silent thread of fixed attention on which the three of us had hung for the last two minutes.

"I told you—I told you it was a beaver! Me and Dad saw him last night—least I was pretty sure it was."

"But Bob!" cried Elisabeth. "He was swimming along with a log . . . well, not a log exactly, but a big chunk of wood in his mouth."

"He's taking it down to his winter storehouse . . . probably a piece of willow. They love the bark."

"But where does he cut it?" put in Miranda. "Last night, remember? I said he looked as if he had something in his mouth, and he was swimming way over by our shoal and heading this way toward Polaris. If he lives on Polaris, why does he go way over to Hay for his wood?"

"Probably no willow on Polaris," I said.

"Then why doesn't he live on Hay where there is willow?" she persisted.

"If we can find him, we'll ask him," I said. Elisabeth was already pulling at the oars, moving the skiff swiftly and silently toward Polaris. We slid along close to the shore, circled the island, but no sign of him.

"Maybe we'll meet him tomorrow night," she said, turning toward home.

"Hope so," said Miranda eagerly. "A beaver, a *real beaver!* I never saw a real beaver up here in the summer! Have you, Mom?"

"Never," replied Elisabeth. "They must be here, but the river's so quiet now that I guess they start moving around openly."

In the chill of late evening, we returned to find the house again full of smoke.

"What I don't understand," said Elisabeth angrily, "I mean if you know what the trouble is, why on earth don't you *do* something about it! Why *don't* you?"

Her point was annoying but well taken. It's no good hooking up two heaters to one six-inch smokepipe. Each heater must have its own exhaust. Nothing hard about grasping that. I had grasped it, or, rather, it had grasped me. The problem consisted in providing the second outlet. There were two ways: 1. Cutting a hole in the roof and running the pipe right up through it; and 2. Knocking another hole in the chimney where it passed through the attic, inserting a pipe, and sealing it in place with cement. Neither seemed attractive.

As for intentionally cutting a hole in one's own roof, that seemed crazy. Besides, the roof already leaked through unfindable cracks where someone else had done the same thing for the flue from a huge exhaust fan over the big gas range. What's more, the people who did it were probably professional roof-cutters. And there were already quite enough buckets for me to tend, delicately balanced on attic joists to catch the drips from their handiwork; so that was no good. The second possibility didn't seem inviting either. The side of the chimney facing the spot where the pipe came up consisted of fieldstone about ten inches thick, and behind that probably would be two courses of brick to go through. The hole through that stone would have to be at least a foot square to give room for working on the bricks behind. Also, the hole would have to be pretty low because it wouldn't be such a good idea having a hot pipe near the rafters. This meant that the knocker—namely, me—would have to maintain a crouch for several hours while swinging a small sledgehammer against a rock chisel. Hours? Days, more likely. Not at all an attractive prospect. The toughest bit would be going through the ten inches of solid stone. One might, instead of actually going through the stone, try simply knocking away the mortar and removing whole stones. That might be more feasible, but then, of course, that could weaken the structure sufficiently for the whole chimney to collapse. So, all things considered, I prayed for fair winds and kept mentioning how I was writing this book.

As September drifted toward a close, the thermometer dipped lower and lower at night, and we had to leave the taps running to keep the water from freezing in the pipes and cracking them. The days grew colder too, and efficient heating became more and more necessary. I was scheduled to fly down to a conference at a college near Baltimore early in October, and as my departure drew closer, my guilt mounted at the thought of leaving Miranda and Elisabeth to maybe freeze alone for five days. Thus driven, on the Friday before my Monday flight, I took sledge, chisel, plus courage in hand and hoisted myself from the top of the stepladder into the attic.

Balanced on the two-inch-wide joist and squatting, I pounded away. Within ten minutes, sweat was dripping off the end of my nose and chin and, judging from the feeling in my legs, it seemed more likely that the South would rise again before I could. With a groan I put down my tools, got a grip on the chimney, and pulled myself upright where I teetered. Grabbing a rafter for support, I kicked out each leg and shook it to loosen the muscles, then bravely squatted again to return to my labors. For the next two minutes I searched in vain for the spot I had been working on. With an agonizing flash came the realization that, with ten minutes of hammering, I hadn't made a dent in the rock big enough even to find. Swallowing that bitter fact, I returned to the attack and stayed at it for the whole of that miserable day.

Around four, Miranda came back from school, and about five, Elisabeth stuck her head through the trap door to say that dinner was almost ready and to ask how I was getting on.

"Not very how," I said, mimicking Eeyore's gloomy response to a question from Pooh.

"Well," she said, "looks from here as though you're trying to go through solid stone!"

"I am," I agreed, creeping along the joists toward the trap door.

"But isn't that crazy? On this side of the chimney facing me, the side where the pipe goes into the chimney, it looks like just brick. Wouldn't it be easier to knock the hole there?"

"Yes," I said sadly, "yes it would. It would be a lot easier, but I don't think the actual flue, the hole inside the chimney, is wide enough to have two six-inch pipes going into it side by side."

"Then put the new one on top of the old one," she suggested brightly.

"I've thought of that too," I moaned, "but I'm afraid of a downdraft pushing the smoke trying to come out the top pipe right back through the bottom one. If it did that, we wouldn't be any better off than we are now."

"Mightn't it do that anyway?" she asked, backing down the ladder to make room for me. "I mean, if you had it in the side? Frankly, I don't see how you can ever get through that rock."

"I agree," I said, climbing down. "I agree with everything. I don't know anything about drafts, but *damn it!* I'm trying!"

That night my dreams were filled with air currents, how they rose, why they descended, but morning found me in the same swirl of uncertainty. Over breakfast I settled their hash—at least in my mind, if not in the real world. I determined to forget the stone, go through the brick, find out what the inside of the chimney looked like, and then decide whether to put the pipes side by side or one on top of the other.

"Oh, yes," I said, "I'm going to need a bag of cement."

"But you don't know anything about cement," Elisabeth countered. "Why don't you call Jack Stuart? I told you to get Jack in the first place!"

Again and most carefully, I explained that it was now Saturday, that I would be going Monday, that I was going to knock a hole in the chimney somewhere or other, and that it was going to take cement to close up that hole; further, that I was not going to be pestering Jack on a Saturday afternoon or a Sunday to do me another favor; that, unless she got the cement—why didn't she want to, anyway?—neither of the heaters would work until I got back, and she and Miranda would be icicles, and so would Ginger, so would Sinbad the cat, and worst of all, so would Ahab the parrot. Therefore, would she please, and so forth, because I was going back to the attic, and I did.

Right away I could tell the difference—after ten minutes of

sledging on the chisel, I could find where I had been whacking. Brick, at least that particular brick, wasn't as hard as the field-stone. After a while, even a few chips flew. I was getting the knack. It consisted of hammering like hell and sharpening the chisel every once in a while. Though hunching there I might be developing a permanent stoop, I might conceivably get through those bricks by evening. Late in the afternoon I heard Elisabeth and Miranda return from town with the groceries.

"Get the cement?" I called down.

"Yes, we got your cement," was the annoyed answer. That was good, because I was going to need it.

"Hey, what are you doing up there? Stuff is coming down the chimney. Great lumps of it."

"Good," I replied. "It's brick and that's because I'm almost through."

"He's going to have the whole chimney down, that's what he's going to do," she said to Miranda. "I don't know why he couldn't wait to get Jack Stuart."

"He's only handsomer," I shouted down; "no smarter than me."

"But he knows a lot more about cement," she returned, "and about chimneys."

Crack! and the last big chunk split off and rattled down into the fireplace along with its own private storm of soot. I was actually *through*. I reached in to get an idea of the size of the hole. It might . . . it actually might be wide enough to lay both pipes side by side. Great! So I began working on the bricks on either side of my breakthrough. They came a lot more easily, because now there was a space into which they could be moved. By dinnertime I imagine I actually did look a "sight," as Elisabeth observed, but there was joy in my heart. My legs, back, and arms were one large ache, and I was covered with a not-so-fine dusting of old mortar, but those two bits of pipe were lying up there side by side in my chimney hole. They weren't connected to anything, so the heaters couldn't be lit, but they did fit. When I cemented them in, hooked up the pipes, and lit those two heaters, we wouldn't have smoke, and we would have heat. It was a good world even if it was a chilly Saturday night. After dinner I asked Miranda if she would guide me with the wheelbarrow while I got some sand from the back bay to mix with the cement.

"It's too late for that, Bob," said Elisabeth. "It's dark, and besides, it doesn't need sand. Least that's what it says on the bag. Mandy, would you get the cement?" A bag of cement weighs ninety pounds. I hadn't thought they would even bother trying to get it up from the boat. Before I could get up to help her, Elisabeth was reading from the bag she held in her hands.

"Let me see that bag," I said.

"Here," she said, casually passing me a paper bag big enough for maybe five pounds of sugar. It was only half full. "It says you add water. That's all."

"You mean this . . . this little bag is . . . the cement?"

"I thought that's what you wanted," she said defensively. "Didn't you want cement?"

"I did. I wanted a *bag* of cement, and . . ."

"Well, that's a bag of cement. What's the matter?"

"The matter is," I moaned, "that there isn't enough cement here to fill a cavity in one of Ginger's back teeth. That hole up there . . . that hole . . ."

"I'm sorry," she said, "but you said you wanted a bag. . . ."

"I know, I know. I didn't say how big a bag, but it only comes in ninety-pound bags. So I didn't think. . . ." She pointed out that it didn't only come in ninety-pound bags, which she could easily prove.

"Unless we can find some cement," I said, "tonight or tomorrow, it's going to be one heck of a cold week around here."

"I bet Jack Stuart's got some," she advised.

"I'm not calling Jack."

"Have it your own way, then. I've done my best. I'm sorry."

Mournfully I took to the back porch and began fighting my way through the pile of junk that protected the other junk in the corners from occasional inquiry of one of my hands or feet. Heaving stuff this way and that, I finally penetrated to a corner where I had a vague memory of having brushed a flap of heavyish paper that could have been . . . yes, was it? Yes, by God! It *was*. It was a half-empty bag of cement. I cheered—too soon, of course, but almost always if you wait around till you've really looked things over, there's seldom anything to cheer about. Anyway, I cheered, and then discovered that the cement was as hard as the fieldstone. Then . . . but then, hadn't I heard stories about how, if you really had to, you could smash pure cement that had gone hard, smash it back to powder, mix it up with sand, and it would harden all over again? What was there to lose? Why not try it, considering the alternative?

Breaking out a few big chunks and putting them into an old tub, I went at them with the sledge, and they broke. It wasn't the sort of thing one might pick out for his favorite pursuit on a Saturday night, but the chunks let themselves gradually be reduced almost to powder.

Miranda cheerfully agreed to go over to the back bay for sand, and with her flashlight she led the way, with me trundling the wheelbarrow behind her, following the sound of her feet.

"You going to mix it tonight, Dad?" she asked when we returned and had sifted out most of the twigs and larger rocks.

"Sure. I'm even going to cement those pipes into the chimney."

"Gosh! Tonight? But that'd be kind of fun. Can I help?" And she did.

"Like making brownies!" she said happily as she sloshed the heavy mixture in a tub with her spade. "Sort of heavy brownies."

"How does it look?" I asked.

"Kind of funny," she admitted.

"Like cement?"

"Well . . . not really, but I don't know what it's supposed to look like." We took turns sloshing, trying to get the water thoroughly mixed in. Every now and again I would ask if it looked any better, and she would point out that she didn't really know what looking better meant, but that it did seem . . . sort of . . . well . . .

As the clock struck ten, I said that that was enough mixing, and we'd go fix those pipes. She got a long extension cord with a light so she could come up too. Together we pushed and patted the mixture around each pipe.

"Pretty messy stuff," she said gleefully, digging out another handful and splotching it into the hole. "Keeps running down where you don't want it, but it's coming on."

Around each pipe I draped a soaking rag and pressed it against the mushy mortar to keep the cement from drying too quickly, and, as the clock in the sitting room struck eleven, we were clambering down the stepladder.

"Mom!" Miranda cried, but there was no answer. She hadn't approved of our carrying on till all hours and had gone to bed.

"Hey, Dad," Miranda asked with happy urgency, "can I have the rest of this cement? Can I?" There was almost half a pail left.

"Sure you can. You want to make something?"

"Yeah, but . . . but I don't know what."

"Be a good idea, actually. We could have a look at it in the morning and see if it hardened." She said she thought cement *always* hardened, wasn't that the whole point? I told her it was, but that this . . . well . . . this wasn't exactly young cement and that Jack Stuart did know lots more than I did about it. She got her flashlight and followed me down to the boathouse where I poured her cement out onto a flat rock, then went down to wash out the two buckets we had used in case the cement should harden. If it hardened quickly, I reflected, I wouldn't be able to get out of my jeans, which were plastered with it. I took them off and washed them out as well as I could, only to find that I was plastered too. Why not? I wondered. Before I could point out to myself that it was much too cold, I was swimming.

"What are you doing!" she cried in disbelief from the shore.

"Having a bit of a wash," I explained. The water was like velvet, almost warm. When I crawled out, pulled on my dripping jeans, and climbed up the stairs and crossed the little bridge, she was completely absorbed.

"Writing *Miranda* with this twig," she said. "When my grandchildren see this . . . I can tell them how . . . how we . . ."

"You'll have to tell them how I went in to get dry and warmed up," and I trudged up the path toward the house. She caught up before I reached it.

"Hurry!" she urged, taking my arm, "let's go in by the . . . by the . . . oh!" and she stopped still. "Stove's not lit . . . not connected, remember?" I remembered. "Anyway, you can dry off and put on your pajamas, and I'll make some hot chocolate."

Dryer and warmer than can be explained, I sipped the hot chocolate at the kitchen table as we reviewed our job. She yawned.

"Time for bed." We walked across to the Sleeping Cabin. As I left her in her room, she added, "Even if it doesn't harden . . . even if it doesn't, we did the job."

"We sure did, Mandy. Thanks. Good night."

Next morning I had the matches in my hand to light the stove before I remembered. Back into my pocket, up into the attic, and across to the chimney. *Miracle!* It was hard. I don't know exactly how hard because I only rapped it with my knuckles, but the pipes were firmly in place and seemed solid enough. I hooked up both heaters and returned to the kitchen, waving matches. In a minute, both heaters were blazing.

"And *no* smoke!" Elisabeth cheered. "Well done." We sat down to a peaceful and leisurely breakfast. Toward the end of the second pot of tea she grew quieter and quieter; then finally, "I don't like to say this. . . ."

"Then don't!" I urged.

"But . . . but . . . there is smoke, and it's coming from the pipe over the stove in the sitting room." I said it couldn't possibly, and she said it jolly well was, and so on and so on.

"Maybe those pipes don't like being side by side," she suggested. "Maybe you should have . . ."

"No. That can't be it."

"Why not?"

"Just *because*. When they were hooked up to the same pipe in the attic, the draft was too weak to drag the soot up. It's probably collected in the first few sections of pipe . . . I hope." Spreading newspaper to protect the "beautiful floor," I brought in the stepladder, climbed up, dismantled the piping from the wood stove, and Elisabeth took each section out and cleaned it. They were dirty.

At the elbow where the pipe turned to go up through the ceiling, I scooped soot out by the happy handful, each one additional proof that I had been right. Suddenly there was a rush of air past my fingers as the ninth-grade-physics "chimney effect" took over. In high glee, I reassembled the scrupulous pipes and relit the heater. It blazed happily. There was plenty of draft—enough so that the wood smoke could not sneak over and come charging back down the space-heater pipe. There would be ample heat during my absence, and *no smoke.*

As soon as I had checked into my hotel in Annapolis, I called Elisabeth to let her know I had arrived safely and to inquire about our smoke problem.

"Not smoking at all," she said. "We're beautiful and warm, but you know what?"

"What?" I asked, putting a hand on the bed for support.

"The pump's quit. We haven't got any water. No water at all; so hurry back!"

After three days of being a panelist and all tied up in heated discussions of Plato, the place of literature, science, and political science in the undergraduate curriculum, I made my way back to the island. By the time I stepped onto the dock, I had figured out what was wrong with the water system and had a clear and simple solution to the problem in my mind. Next morning, wrenches in hand, I marched purposefully to the pumphouse to take care of things. It took thirty seconds to puncture my carefully inflated balloon. I had been absolutely wrong in my diagnosis, so my beautiful solution had to be discarded like a piece of old Kleenex. Still, it only took us eight hours of joint hard labor to find out what the trouble really was and to get things functioning properly so that by the next morning we were back in our happy routine.

Brr! at 7 A.M. the floor of the Master Cabin is like ice under the bare feet.

"Miranda! Miranda! Time to get up!" She groans her sleepy acknowledgment as, slippered, I go through her room to the door facing the Main House. Ginger yawns, testing the morning with two or three tentative licks of her tongue, thumps down from Miranda's bed, and waddles out after me. Ever delicate in such matters, she pretends she "doesn't need to go"; sniffing casually about in the leaves, she lets it be known there is something around the back, or perhaps off in those bushes—there is something in that general direction that needs attending to . . . and if I'll just go ahead and see to the fire, she'll be along soon. I open the door into the sitting room where the clock has conscientiously ticked through the night and where the wood-burning stove has been dutifully fighting off the cold.

Through the sitting room into the big kitchen. Fill the kettle, put it on the huge bottle-gas range and twist the valve. *Boof!* as the flame sizzles moisture from the bottom of the kettle. Then turn on the radio through which the disc jockey, sitting in his warm studio in Kingston twenty miles upriver, announces that the temperature is twenty-five degrees and that yesterday the Catterockway Raiders defeated the Napanie Bombers three to one. The boyng-boyng of electric guitars and the cries and moans of a tortured, I guess, vocalist fill the empty kitchen as I return to the sitting room to shake down the ashes and carefully rebuild the fire. Muffled in robe and slippers, Elisabeth comes in with a sleepy, "Hello. Cold morning, eh? Did the fires stay in?" On, then, into the kitchen where she starts the porridge, Red River cereal, or maybe kippers. Then Miranda's drowsy, "Hello," bending for a moment over the stove for warmth, then into the kitchen to set the table. We breakfast with the river in full view through our new six- by three-foot plate-glass window.

"It's so lovely," says Elisabeth, "I don't even miss reading the morning paper."

With the eight-o'clock news, Miranda goes back to the Master Cabin to finish dressing, and, clutching a cup of hot tea, I go down to the boathouse to bail out her boat, make sure there's enough gas in the tank, and finally start her motor.

"Okay, Dad! Thanks!" she calls, tossing in the lines and stepping in as I climb out to stand beside Elisabeth. Putting the motor in reverse, she backs slowly out of the slip, and, waving our cups of tea, we shout our good-bys to her as she swings around, turns up the throttle, and buzzes off toward town and her seventh-grade classroom, which is a fifteen-minute walk from the Bay Street Dock.

"Good-by," she shouts back to us as the boat drones away. In silence, we stand—Elisabeth watching, I listening—till she turns the corner of Foresyth, three quarters of a mile away, and into the last straight stretch to shore.

"She won't forget these mornings," observed Elisabeth as we climbed the steps beside the boathouse and made our way toward the quiet Main House.

"Hope not," I answered.

"Aren't you cold? Come on. Let's hurry back for another cup of tea!" We dawdle till nine or even later over the pot till she exclaims, "Disgraceful! This will never do. We've got to get on with the day! Let's have our baths."

When we return from the Master Cabin, the heater in the sitting room is blazing. Adjusting its draft, I settle down at the typewriter. With swift efficiency, she cleans the kitchen and, if the sun has come out, she says, "Think it's warm enough to paint? Says on the can it's no good below fifty degrees, and there are still those windows to do on James's cabin," and she is off.

After lunch, perhaps we take the chainsaw to cut up a fallen oak and trundle the logs back in the wheelbarrow to where I split them. Around four, Ginger's excited barking announces that Miranda is pulling in from school, so we go down to meet her.

"And did you have a good day?" we ask.

"Yep," she answers, tossing up a bowline and cutting the motor.

"What did you do?"

"Nothing," comes the happy reply. "We won our soccer game."

"Hurray!"

"What's more, I scored the winning goal, and they say I might make the all-star team!"

"Great!" giving her a hug.

"Yep. Is there any tea?"

"I'll put the kettle on," says Elisabeth. "How about you running over to check Carpenter's?"

"Okay. Come on, Ginger. Let's go," and together they trot happily off through the trees. Carpenter's cottage is in a bay on the other side of the island, about a fifteen-minute tramp from the boathouse. Cliff had hired Miranda as his winter caretaker, and she went over every day to see that nothing had been disturbed. She enjoyed earning the money, but she enjoyed the walk even more. One day when I walked over with her I casually asked, "You ever get scared or lonely coming over here all by yourself?"

"Oh, no! It's lovely just going along through the trees with the island all empty like this and nobody around. I can pretend I'm an Indian, you know, just stalking along, or maybe even a leopard! Anything—I can be anything I like. I creep along by the back bay where the path goes through a whole bunch of trees, and it's all sort of gloomy and kind of dark and mysterious. And then maybe I suddenly jump off that big rock out onto the beach where everything is open and pretend . . . well, pretend anything! anything at all!

"So I don't get scared at all. Oh, it was nice when Shelley or somebody would come with me, but then, you know, I couldn't do things like just sit down and look at a plant if I wanted to or maybe yell to hear the echoes. It's kind o' . . . I don't know . . . embarrassing or silly or something, with other people there, so you don't do it. That's why I like it better, really, going by myself. Oh, I'm glad to have Ginger, but she's different. She doesn't care what you do, and you can tell her anything, and she listens and doesn't mind. You can tell her absolutely anything, and she just wags her tail, licks your hand, and is all gay and happy. Then she dashes off to chase a rabbit or something . . . clumsy old fat thing, you know, thumping and crashing along through the bushes . . . and the rabbit skips away laughing, but Ginger doesn't mind. She doesn't mind not catching them. She comes back panting and happy and all excited. Then I chase her for a bit; then she chases me. She's great to have along.

"Only thing a bit scarey is the house. I don't like going up onto the porch to try the doorknobs. I do that real fast. Then I run back down onto the grass. I don't know what I expect to happen —I guess it's that the house is empty. I mean, I know that. It really is empty, but, you know, people have been there. Funny, really . . . the island . . . well, the land isn't like that! It can be empty, and even though you know people have been there, it somehow isn't scarey like a house. But I kind of feel the house is sort of . . . sounds strange . . . watching me . . . just watching. . . . Not that it would actually *do* anything to me, you know, but it's just quietly watching, and that makes me feel kind of funny, so I guess I don't like that! Me and Ginger run back to the beach where everything is all open and clear, and we start back.

"It's lovely to start back, too, because I know the heaters are going and that everything'll be all snug and warm, and there'll be either nice hot tea or supper waiting. Then after supper there are the stories, and I can hardly wait to find out what's going to happen next, so coming home is lovely too."

The stories are six novels of Dickens that we rolled through together. They were either on talking-book records or magnetic tape, "recorded," as the reader announced firmly at the beginning of each book, "solely for the use of the blind or physically handicapped."

"Gee," said Miranda at the beginning of the first one, "do you think it's all right, then, for me and Mom to listen too?"

"If we keep very quiet," said Elisabeth, "we probably won't be noticed."

As the reader spun through chapter after chapter, I tended the heater and Elisabeth's knitting needles fairly purred while Miranda's clicked occasionally with the desperate concentration of a beginner. Dickens welded those cozy evenings together into one long stream of suspense, amusement, and delight while the television set sulked gloomily in the chill silence of the upper part of the boathouse. After starting to yawn around nine-thirty, Miranda would finally declare about ten, "Can you turn it off, Dad? I've got to get to bed."

"Sure. Sure," I answered guiltily, "soon as this record's finished."

"Yeah, but Dad! If that was the TV and I wanted to watch to the end of the program, what would *you* say? Huh? What would you *say?*"

"Okay," I mimicked her Lancaster TV voice, "just a minute." She laughed sleepily as I turned off the machine. Grabbing up the cat and calling Ginger, she opened the door and marched out with flashlight in hand to cross the hundred feet of blackness to the Master Cabin.

Elisabeth put away her knitting and I bedded down the fire, covered the parrot, then got ready myself.

"Well," she said, stopping at the door, "this sitting room has been a success, hasn't it?"

"And so have lots of other things," I said, and we followed Miranda.

19

Autumn

It was early October. In the last five weeks the hum of river traffic had slowly died till the sound of a passing boat pulled Elisabeth to a window to see whose it was and where it was headed. As the humans gradually withdrew, animals and birds cautiously advanced to repossess their territories. The occasional squirrel, for example, that Ginger pursued with sudden fury in August, had miraculously multiplied into families, tribes, whole colonies. By October Ginger would sit whining and panting with frustration, but unmoving because she couldn't make up her mind which one or ones to chase. Every sunrise seemed to bring forth still another band, every member of which was bent on laying in a full store of winter supplies. No shilly-shallying about in the mornings for them. They helped crack dawn with their squabblings over high clusters of pine cones and acorns, and they brought storms of both raining like hail upon cabin roofs. The grasshopper time of summer had passed with its morning chorus of song sparrows, robins, whitethroats, and orioles—theirs had been a world of fun and games—but this was the season of the ant—all business—and the squirrels went about it in full seriousness. And while at work, they sent up to heaven no chorus of shallow, pastoral joy. They raised their voices in anger, outrage, despair, melancholy, anguish, jeers, taunts, jibes, recriminations—all expressing their deep-seated conviction that, it being autumn, life was no laughing or singing matter. They agreed with the cawing crows that it was tough, that it was going shortly to get a lot tougher, and that, unless one did something very definite and very hard, and very soon, it could be a mortal affair. Of course, the crows had been saying this all summer, but who, except a crow, could hold to such a gloomy conviction through those blue and silver months, full of warmth and joy?

As the squirrels appeared from nowhere to shriek about what was to come, from nowhere, too, came owls in the evenings to hoot their ghostly warnings, that the squirrels were right. Earlier and earlier they came, flitting, as it were, from the darkness of night into the indistinct twilights, then even forward into the lands of afternoon to send their quivering, eerie voices through a quieter and quieter country. So far did they intrude into daylight that the ever-watchful crows began to notice and to object. Like dark packs of yelping hounds, they set upon the invader, to surround, to fling themselves at him, to drive him back into the darkness from whence he came, as if they saw in him the figure of Death itself. Screaming, they hurl themselves against the razor-beaked, steel-taloned apparition, and through their violence publish the pathos of their contradiction: "Yes. We knew about you . . . believed in you . . . said we believed all through the sweetness of summer . . . but your time is not yet! Not for us . . . Not now! Back to . . . back to darkness."

And, far off, the single crow, winging through an empty autumn sky, hears and recognizes the clamor. With a hoarse cry he wheels and flings himself along to join the desperate struggle. Beating furiously toward the combat, he blazes out, "I'm coming! Don't give up! I will help! What the geese won months ago, we shall defend. He shall not return!" And he, too, disappears into the boiling turmoil of sound and fury of the not-long-since silent pine in whose branches now crouches the embattled owl.

And they win—they do win. . . . This afternoon, at least, they seem to prevail; for, after five, ten, perhaps fifteen minutes, the gray phantom, conceding that his time is not quite yet—admitting, perhaps, he has been a trifle previous—sails from his perch to find refuge in some denser grove where his black tormentors cannot maneuver quite so freely, and there to wait; but not for long now, because his hour is near at hand.

One morning late in the fall we rose to find the river already in high glee and bannered with whitecaps saluting a proud southwesterly wind. This was no morning for Miranda to cross in her little twelve-foot boat; so we fell back upon our seventeen-foot Chris Craft, a mighty inboard-outboard. *Chris Craft!* surely a name to conjure with—or so I had thought. And, so thinking, I had feverishly snatched this great buy at the end of the previous year. Only afterward did I discover that my unerring nose had scented out in this 1964 model Chris Craft's reverse Edsel. Though visually an atrocity, that car, I am told, worked like a charm; whereas the 1964 Chris Craft looked like a dream, but ran, if at all, like a nightmare. That high morning it eagerly took us halfway to town, then resigned and left us to our desperate paddles. It was only with the help of a good Samaritan on shore that we narrowly escaped disaster. At last, the glorious hull with its rotten engine moored safely at the Bay Street Dock, Miranda on her way to school, the telephone call for help having been made, Elisabeth and I waited on the dock for the mechanic.

After half an hour a car pulled in and stopped in the parking lot. But it wasn't the mechanic. The fellow came out onto the dock and said good morning. He wanted to know if we had seen a guide boat around. He was going fishing and hoped he hadn't missed his guide. We told him there had not been a boat in the last half hour. He relaxed and chatted. He said he was from around Scranton, Pennsylvania. We told him we were from Lancaster, and that pleased him.

"Imagine that!" he said. "I know that whole section down through there. Used to deliver coal down around there. Good farming country. You folks had a farm, did you?" We said we hadn't. "Well," he said with a deep sigh, "well . . . it's good to see you *both.* I just buried my wife. Funeral was day before yesterday."

"Sorry," we said.

"Yes," he said thoughtfully, "day before yesterday. Alone now." Then: "Worked hard all my life, I have, and so did she . . . so did she. We had six children. Then my brother died and we took in his four. That woman, she worked from dawn till dark —cooking, scrubbing. Worked her like a slave, I did . . . just like a slave. Mind you, I worked hard myself. Business, business; always business. Sometimes be twelve, one o'clock before I got home, and then come the weekends, something always turned up I had to do. And there she was with all those kids.

"I'd say, 'Come on, let's go out to dinner or a show, or a weekend. Everybody needs a vacation sometimes,' but she'd always say no, that when the kids were gone, then we'd have all the time in the world! That's what she used to say!" He gave a short, bitter laugh. "And I said, 'Sure, okay,' because the kids, they needed a good home, and she gave 'em a good home. And me, I wanted to build a good business so they could have things.

"Now, let me tell you something! I got a good business, see—a fine business! We could go anywhere in the world now, anywhere! The kids are gone—anywhere in the world! Yeah!" And with a rasp of sadness, "And so is she. So is she." Wind and wave hissed and rippled through the silence of his pause. He collected himself slowly for strong speech.

"Let me tell you—both of you—I didn't have much education, but I learned one thing. And I tell you what I learned. Don't wait!" He spat into the river. "You wait, and what for? Alone. What good am I? What good's my business . . . that great business. All those years! She never even been up here. She never been anywhere. You listen. I know. Don't wait for something, because maybe it don't come, see? Maybe never; so then what's it all for, I'd like to know? Tell me *that!*

"Ah! Here comes Stan; so I go fishing, but you remember. It's no shame to learn from an old man, is it? I wish an old man had talked to me. Hi, Stan. Be pretty rough day, eh? Good-by folks. Have a good time." And the St. Lawrence Charon, that ancient ferryman for the dead in Greek mythology, took his passenger out into the hissing, rippling emptiness of autumn, leaving us on the edge of a suddenly darker stream. We moved closer to each other, and Elisabeth took my hand. After a while, she said, "It isn't going to be like that for one of us someday, is it, darling?"

"No," I answered, "not like that. Whatever else is ahead for us, it won't be that. We didn't wait, you know. I'm not quite sure why we didn't, but we didn't, so we have . . . all of us have had the whole web of fifteen summers. They've tied us all together . . . tied us all to this, and none of us can ever be sorry . . . not about *that!*"